Praise for *Good Questions for Math Teaching,*

Questioning lies at the heart of good instru
coach, teacher, and teacher educator should have a copy of

—*Mary Kay Stein, Professor, University of Pittsburgh, Pennsylvania*

Good Questions has been an essential tool for supporting remote teaching. This resource is one I will use and keep forever; I highly recommend it to current and future teachers.

—*Rachel McAuliffe, third-grade teacher, Victoria, Australia*

Math teachers are always on the lookout for great questions to ask their students—questions that offer the right amount of challenge, that can be approached in multiple ways, and that allow teachers the opportunity to assess what students know. *Good Questions* is a wonderful resource for exactly these kinds of questions. *Good Questions* is packed with excellent questions from a wide range of K–5 content domains and can be used in conjunction with any curriculum. It is definitely worth a look!

—*Jon R. Star, Professor of Education, Harvard Graduate School of Education, Boston, Massachusetts*

This series provides excellent guidance for opening up questions, with a very clear explanation of the four main features of good questions and many immediately usable examples. New teachers will use this resource to learn what good questions are, and veteran teachers will use it to create and refine their own good questions. I use it in my work with future teachers as well as K–5 students!

—*Julie McNamara, author of* Beyond Pizzas and Pies: 10 Essential Strategies for Supporting Fraction Sense, Second Edition

As a vocational adult education trainer and assessor, I have found the *Good Questions* series to be an invaluable resource for developing the problem-solving skills of supervisors and managers. *Good Questions* provides excellent examples of the questioning skills that enable supervisors and managers to be confident that their staff and teams actually understand the underlying requirements of the required task.

—*Mike Stoll, MBA, Dip Training and Assessment, Victoria, Australia*

All teachers ask questions; however, *Good Questions* provides them with open-ended questions that will spark discussion and will force students to think deeply about the mathematics. All discussions start with a good question, and this book has plenty!

—*Suzanne H. Chapin, coauthor of* Talk Moves: A Teacher's Guide for Using Classroom Discussions in Math, Third Edition

GRADES K–5

GOOD QUESTIONS FOR MATH TEACHING

WHY ASK THEM AND WHAT TO ASK

SECOND EDITION

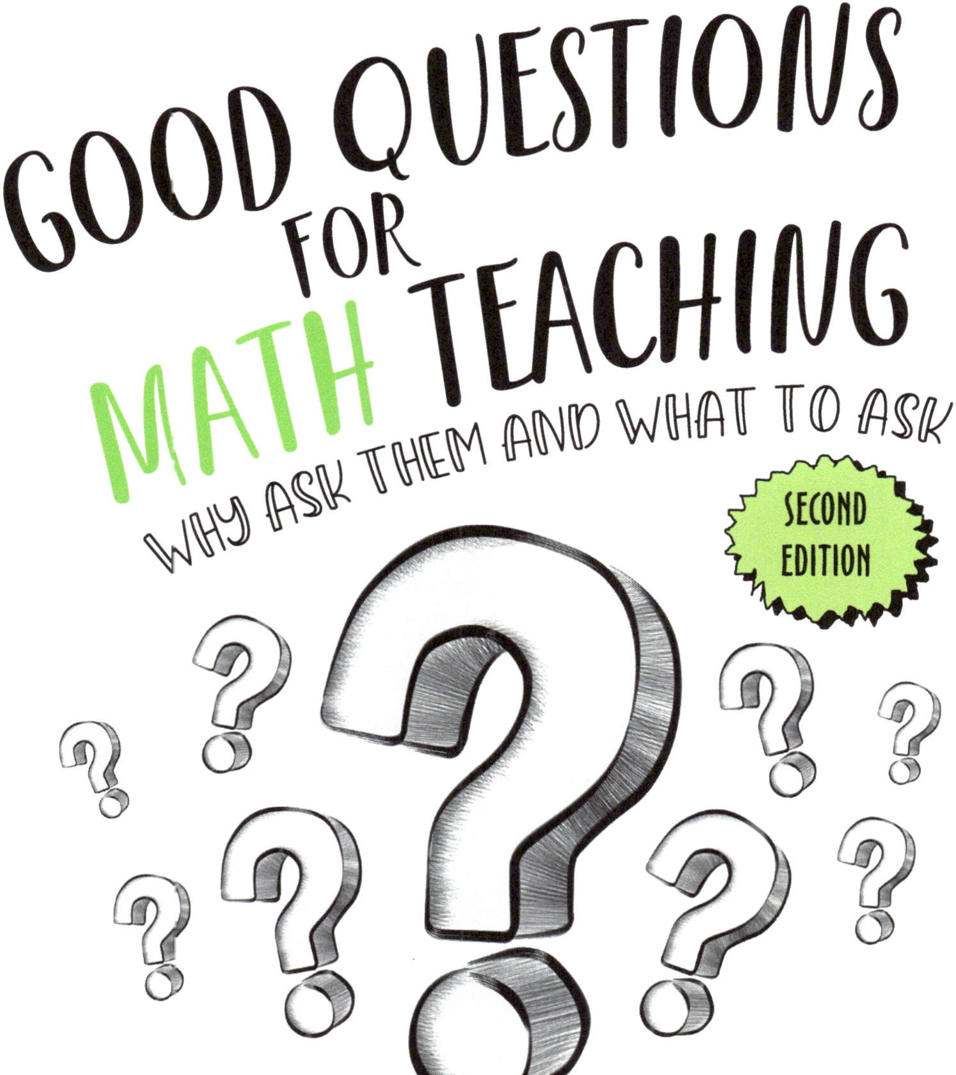

PETER SULLIVAN · PAT LILBURN

Series Editor, Nancy Anderson

HEINEMANN

Portsmouth, NH

Heinemann
145 Maplewood Avenue, Suite 300
Portsmouth, NH 03801
www.heinemann.com

Good Questions for Math Teaching, Grades K–5, Second Edition was originally published by Houghton Mifflin Harcourt under the Math Solutions brand, ISBN: 978-1-935099-76-5.

Cataloging-in-Publication data is on file with the Library of Congress.

ISBN-13: 978-0-325-13759-9
eISBN-13: 978-0-325-13940-1

Executive Editor: Jamie Ann Cross
Production Manager: Denise A. Botelho
Editorial Assistant: Kirby Sandmeyer
Photo Credits: Author photos of Peter Sullivan and Pat Lilburn provided by the authors.
Interior Design and Composition: Publishers' Design and Production Services, Inc.
Cover Design: Susan Olinsky Design
Manufacturing: Gerard Clancy
Cover Art: Question mark art created by sdecoret / Shutterstock

Printed in the United States of America on acid-free paper.
2 3 4 5 GP 26 25 24 23 PO4500871421

A Message from Heinemann

Heinemann's math professional resources are written by educators, for educators, to support student-centered teaching and learning. Our authors provide classroom-tested guidance, advice, and proven best practices to help teachers increase their comfort and confidence with teaching math. We believe a focus on reasoning and understanding is the pathway to helping students make sense of the mathematics they're learning.

This resource was originally published by Math Solutions, a company long dedicated to similar ideals and aims as Heinemann. In 2022, Math Solutions Publications became part of Heinemann. While the logo on the cover is different, the heart of Math Solutions lives in these pages: that teaching math well calls for increasing our understanding of the math we teach, seeking deeper insights into how students learn mathematics, and refining our lessons to best promote students' learning.

To learn more about our resources and authors, please visit Heinemann.com/Math.

Brief Contents

List of Reproducibles xiii

Acknowledgments xv

How to Use This Resource xvii

Why This Resource? xvii

How Is This Resource Organized? xviii

When Should I Use These Questions? xviii

PART I The Importance of Good Questions 1

CHAPTER 1 What Are Good Questions? 3

CHAPTER 2 How Do I Create Good Questions? 9

CHAPTER 3 Five Steps to Successfully Implementing Good Questions in Your Classroom 13

PART II Good Questions to Use in Math Lessons 21

CHAPTER 4 Good Questions for Money and Financial Mathematics 23

CHAPTER 5 Good Questions for Counting and Place Value 35

CHAPTER 6 Good Questions for Decimals 53

CHAPTER 7 Good Questions for Operations 65

(Continued)

CHAPTER 8	Good Questions for Fractions	81
CHAPTER 9	Good Questions for Geometry	93
CHAPTER 10	Good Questions for Data Analysis and Probability	115
CHAPTER 11	Good Questions for Measurement	135

Reproducibles 173

References 203

How to Access Online Resources

To access the downloadable reproducibles referenced in the text, please see page xx for directions, visit http://hein.pub/MathOLR, and register for an account. Use key code **GQK5** to register this product and download.

Contents

List of Reproducibles		xiii
Acknowledgments		xv
How to Use This Resource		xvii
Why This Resource?		xvii
How Is This Resource Organized?		xviii
When Should I Use These Questions?		xviii
PART I	**The Importance of Good Questions**	**1**
CHAPTER 1	What Are Good Questions?	3
CHAPTER 2	How Do I Create Good Questions?	9
CHAPTER 3	Five Steps to Successfully Implementing Good Questions in Your Classroom	13
PART II	**Good Questions to Use in Math Lessons**	**21**
CHAPTER 4	Good Questions for Money and Financial Mathematics	23
	Grades K–2	24
	Money	*24*
	Grades 3–5	30
	Money	*30*

(Continued)

CHAPTER 5 Good Questions for Counting and Place Value 35
Grades K–2 36
Counting *36*
Place Value *40*
Grades 3–5 44
Counting *44*
Place Value *48*

CHAPTER 6 Good Questions for Decimals 53
Grades 3–5 54
Decimal Concepts *54*

CHAPTER 7 Good Questions for Operations 65
Grades K–2 66
Addition and Subtraction *66*
Grades 3–5 72
Addition and Subtraction, Multiplication and Division *72*

CHAPTER 8 Good Questions for Fractions 81
Grades K–2 81
Fraction Concepts *82*
Grades 3–5 86
Fraction Models, Comparing Fractions, Adding and Subtracting Fractions *86*

CHAPTER 9 Good Questions for Geometry 93
Grades K–2 94
Two-Dimensional Shapes *94*
Three-Dimensional Shapes *98*
Grades 3–5 104
Two-Dimensional Shapes *104*
Three-Dimensional Shapes *111*

CHAPTER 10 Good Questions for Data Analysis and Probability 115
Grades K–2 116
Chance *116*
Data *120*
Grades 3–5 123
Chance *123*
Data *127*

CHAPTER 11 Good Questions for Measurement 135
Grades K–2 136
Weight 136
Time 140
Length and Perimeter 145
Grades 3–5 149
Weight 149
Volume 152
Area 158
Time 165
Length and Perimeter 168

Reproducibles 173

References 203

How to Access Online Resources

To access the downloadable reproducibles referenced in the text, please see page xx for directions, visit http://hein.pub/MathOLR, and register for an account. Use key code **GQK5** to register this product and download the reproducibles.

Reproducibles

The following reproducibles are referenced throughout the text. These reproducibles are also available in a downloadable, printable format. For access, visit http://hein.pub/MathOLR and register your product using the key code **GQK5**. See page xx for more detailed instructions.

Reproducible 1	Cafeteria Price List
Reproducible 2	Money Number Cloud
Reproducible 3	Jigsaw Piece
Reproducible 4	Decimal Number Cloud
Reproducible 5	Triangles
Reproducible 6	Mixed Two-Dimensional Shapes
Reproducible 7	Mixed Three-Dimensional Shapes
Reproducible 8	Walking Path Shape
Reproducible 9	Block Diagram
Reproducible 10	Mystery Bar Graph
Reproducible 11	Mystery Line Plot
Reproducible 12	Blank Spinner Faces

(Continued)

Reproducible 13 Mystery Pictograph

Reproducible 14 Favorite Shows Pie Chart

Reproducible 15 Hunger Graph

Reproducible 16 Mystery Survey Results

Reproducible 17 Mystery Pie Chart

Reproducible 18 Sports Pie Chart

Reproducible 19 Animal Pictograph

Reproducible 20 Measurement Line Plot

Reproducible 21 Graph of Children Talking

Reproducible 22 Cubic Structure

Reproducible 23 Rectangular Box

Reproducible 24 Cube Diagram

Reproducible 25 Letter O

Reproducible 26 Package with Ribbon

Reproducible A Hundreds Chart

Acknowledgments

The idea of open-ended or good questions developed over several years during ongoing discussions between Peter Sullivan and David Clarke.

Peter and David conducted a number of research studies and classroom trials of open-ended questions. Many of David's initial ideas are used in various places throughout this book. His creativity, energy, and interest in exploring good questions contributed significantly to the idea of using open-ended activities in the teaching of mathematics, and for this we thank him.

The idea went through a number of phases before it reached its final form. Pam Rawson's contribution to the early planning stages, which ultimately led to the development of this resource, is greatly appreciated.

Finally, we thank Sheryl and Mike, without whose support there would have been no book.

—Peter Sullivan and Pat Lilburn

How to Use This Resource

Why This Resource?

Our goals of education are for our students to think, to learn, to analyze, to criticize, and to be able to solve unfamiliar problems, and it follows that good questions should be part of the instructional repertoire of all teachers of mathematics. As we work to emphasize problem solving, the application of concepts and procedures, and the development of a variety of thinking skills in our mathematics curricula, it becomes vital that we pay increased attention to the improvement of our questioning techniques in mathematics lessons. As teachers of mathematics, we want our students not only to understand what they think but also to be able to articulate how they arrived at those understandings. Developing productive questions can help focus learning on the process of thinking while attending to the study of content (Dantonio and Beisenherz 2001, 60). Good questions created and posed by teachers ultimately become powerful tools for student learning.

In *Good Questions for Math Teaching, Grades K–5, Second Edition*, we describe the features of good questions, show how to create good questions, give some practical ideas for using them in your classroom, and provide many good questions that you can use in your mathematics program. By asking careful, purposeful questions, teachers create dynamic learning environments, help students make sense of math, and unravel misconceptions.

How Is This Resource Organized?

This resource is divided into two parts: "Part I: The Importance of Good Questions" and "Part II: Good Questions to Use in Math Lessons."

Part I

In Part I, we explain what we mean by "good questions," walk you through different methods to create your own good questions, and provide necessary support for how to use good questions in your classroom.

Part II

In Part II, we offer examples of good questions, broken down by topic and grade span, for you to select from and use in your classroom. Each question is accompanied by notes to support your teaching.

When Should I Use These Questions?

Here are a few of our favorite scenarios for when to use good questions.

> **When to Use Good Questions**
>
> - Use good questions to supplement your mathematics curriculum.
> - Use good questions for warm-up routines or for review.
> - Incorporate good questions into homework and/or assessments.

Use Good Questions to Supplement Your Mathematics Curriculum

The questions in this book are designed as a supplement to your mathematics curriculum. Good questions can be used as the basis for an entire lesson, either as a lesson that stands alone or as part of a unit of study. As you progress through a unit of study, you may want to ask your students questions in this book that correspond to that unit. Embedding questions from this book within your lessons may further enhance student learning and understanding. When studying subtraction, for example, you may wish to refer to the questions in this book to support, extend, or enrich your present curriculum. If current practice has your students solving word problems that have only one answer, you may wish to ask students questions that allow for multiple answers—and approaches—such as Chapter 7, Question 16 on page 71 (*I have some marbles. I give some away to my friends and am left with fifteen. How many marbles might I have started with and how many might I have given away?*). By constructing an answer to this question, students gain valuable experience with subtraction and also gain insights into the relationship between addition and subtraction as two operations that "undo" one another.

Use Good Questions for Warm-Up Routines or for Review

Another way to use this book is to use the questions as a daily warm-up activity or "do now" activity for the start of math class. You may choose a question that corresponds to the current unit or a question that requires students to review a particularly challenging or important skill or concept. Students can talk about the question with their peers before you convene the class with a discussion of students' answers. This routine makes good use of transition time while immediately focusing math class on reasoning and communication.

Incorporate Good Questions into Homework and/or Assessments

In addition to using these questions during instructional time, you could assign them for homework or incorporate them in your assessments. Typically, math homework and assessment practices tend to focus on skills and/or closed questions that require recall of what was learned in class. If your students usually complete a review worksheet of the day's lesson, attach (or even substitute) a question from this book that will require them to think beyond what they have learned, to connect an understanding from a previous lesson, or to confront a misconception that may have arisen during class. In addition, if students' assessments normally include computations and problems for which there is only one correct answer, adapt those assessments to include some questions from this book that will allow students to think creatively about the mathematics they are learning.

How to Access Online Resources

1. Go to http://hein.pub/MathOLR and log in if you already have an account. If you do not, click or tap the Create New Account button at the bottom of the Log In form.
2. Create an account. You will receive a confirmation email when your account has been created.
3. Once your account has been created, you will be taken to the Product Registration page. Click Register on the product you would like to access (in this case, *Good Questions for Math Teaching, Grades K–5, Second Edition*).
4. Enter key code **GQK5** and click or tap the Submit Key Code button.
5. Click or tap the Complete Registration button.
6. To access the reproducibles at any time, visit your account page.

[Key Code GQK5]

More Resources in This Series!

We are excited to share this resource, *Good Questions for Math Teaching, Grades K–5, Second Edition*, as one of several resources in the Good Questions series. For more titles, see Heinemann.com/Math.

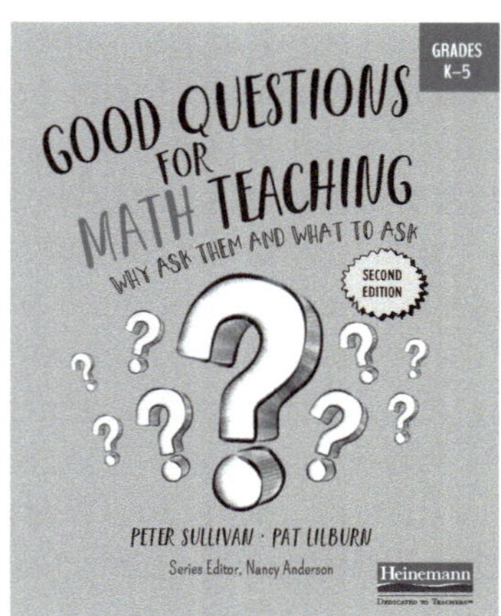

Grades K–5
ISBN: 978-0-325-13759-9

Grades 5–8
ISBN: 978-0-325-13760-5

High School
ISBN: 978-0-325-16026-9

PART I

The Importance of Good Questions

CHAPTER 1	What Are Good Questions?	3
CHAPTER 2	How Do I Create Good Questions?	9
CHAPTER 3	Five Steps to Successfully Implementing Good Questions in Your Classroom	13

In Part I, we explain what we mean by good questions, walk you through different methods to create your own good questions, and provide necessary support for how to use good questions in your classroom. Remember, the power of questioning is ultimately in the answering. As teachers, we not only need to ask good questions to get good answers but need to ask good questions to promote the thinking required to give good answers.

Chapter 1

What Are Good Questions?

During the course of a normal school day teachers ask many questions. In fact, something like 60 percent of the things said by teachers are questions and most of these are not planned. One way of categorizing questions is to describe them as either open or closed. Closed questions are those that simply require an answer or a response to be given from memory, such as a description of a situation or object or the reproduction of a skill. Open questions are those that require a student to think more deeply and to give a response that involves more than recalling a fact or reproducing a skill.

Teachers are usually skilled at asking open questions in content areas such as language arts or social studies. For example, teachers often ask children to interpret situations or justify opinions. However, in mathematics lessons closed questions are much more common.

Questions that encourage students to do more than recall known facts have the potential to stimulate thinking and reasoning. To emphasize problem solving, application, and the development of a variety of thinking skills it is vital that we pay more attention to improving our questioning in mathematics lessons. Teachers should use questions that develop their students' higher levels of thinking.

Good Questions for Math Teaching, Grades K–5, looks in more detail at a particular type of open question that we call a "good" question. Good questions can set the stage for meaningful classroom discussion and learning. When we ask good questions in math class, we invite our students to think, to understand, and to share a mathematical journey with their classmates and teachers alike. Students are no longer passive receivers of information when asked questions that challenge their understandings and convictions about mathematics. They become active and engaged in the construction of their own mathematical understanding and knowledge.

Careful, intentional, and mindful questioning is one of the most powerful tools a skillful teacher possesses (Costa and Kallick 2000,

34). So, what do good questions "look" like? Let's consider four main features.

Four Main Features of Good Questions

1. Good questions help students make sense of the mathematics.
2. Students can learn by answering good questions, and the teacher learns about each student from the attempt.
3. Good questions may have several acceptable answers.
4. Good questions are accessible to all students.

Feature 1: Good Questions Help Students Make Sense of the Mathematics

Too often mathematics is presented in a closed or rote manner: Students are given a procedure (or told about a concept) and then asked to absorb that information without much reflection or exploration. Good questions, on the other hand, require more than remembering a fact or reproducing a skill. Their open-endedness requires the application of facts and procedures while simultaneously encouraging students to make connections and generalizations. Good questions help students see mathematics as a subject that is based on sense making. In turn, good questions empower students to think of themselves as capable sense makers. Let's look at several scenarios to help further explore this.

Scenario 1: Place-Value Questions

Think about the following two questions—which one do you feel is more powerful in helping a student make sense of mathematics? Why?

> **Question A:** *What digit is in the tens place of the number 63? What is its value?*
>
> **Question B:** *I am thinking of a number between 10 and 100 with a single 9 in it. What might my number be?*

Question A might be considered a typical kind of place-value question that is asked of students in the early elementary grades. These questions typically present students with a number and ask them to identify each digit by its face and place value. While we would not want these questions to disappear entirely, we also recognize that these questions have their limitations. Specifically, they do not empower students to make decisions based on what makes sense or consider connections between important mathematical concepts. What if we took Question A and opened it up? That's where Question B—what we consider a "good question"—comes into play. Question B invites students to dig deeply into the relationship between the value of a digit and its position in a number while also making decisions about how to organize their thinking, eliminate possibilities, and find all possible answers. By opening up the question, we empower students to think of themselves as capable sense makers. When students are given many opportunities to derive their own strategies, try them out, make adjustments and justify their reasoning, they gain knowledge, thinking skills, and confidence. And when students are exposed to a classroom environment where good questions are the norm, something else wonderful happens too. The "let's try it and see what happens" thinking at the heart of any good question also motivates students to generate their own good questions! In Question B, since there is a range of acceptable answers (e.g., 19, 49, 91), students might wonder:

- What is the total number of possibilities?
- What is the largest number possible? The smallest number possible?
- How might my strategy or approach change if the givens of the problem change—say, the number was between 10 and 200?

The nature of the question stimulates students' natural sense of curiosity, which in turn fosters higher-level thinking and the problem-solving skills that are essential to the work of doing mathematics in school and beyond.

Scenario 2: Averages

Let's look at another example. Consider these three questions—which one(s) invite students to make sense of mathematics?

Question A: *The average of five numbers is 6. What might the numbers be?*

Question B: *What is the average of 6, 7, 5, 8, and 4?*

Question C: *After five games, the goalie had averaged blocking six goals per game. What might be the number of goals she blocked in each game?*

Question B is another example of a closed question commonly found in textbooks. This question mainly requires students to recall a technique. That is, add the numbers and divide by how many there are—in this case five. Questions A and C, on the other hand, are an example of opening the question up—rephrasing it in the form of a good question. Both Question A and Question C require a different level of thinking and a different type of understanding of the topic of averages to be able to give an answer. Students need to comprehend and analyze the task. They must have a clear idea of the concept of average and either use the principle that the scores are evenly placed about the average or that the total of the scores is 30 (that is, 5 × 6) as the basis of their response. It most definitely requires more than remembering.

Scenario 3: Operations

Here's one more example of the power of a good question. MJ, a second grader, had just finished a unit on addition and subtraction where she had been asked to model story problems with number sentences. She was able to complete most of these correctly, and the teacher assumed from this that MJ could use the actions and relationships in the story problem to choose the appropriate operation. With this in mind, the teacher gave MJ the following good question:

I have some marbles. I give some away to my friends and am left with fifteen. How many marbles might I have started with and how many might I have given away?

MJ read the question several times, and exasperated, responded, "I can't do this! There isn't enough information." What happened?

To find an answer, MJ needed to think about the relationship between addition and subtraction. She had to consider that although the story problem describes a "take away" or "removal" situation, it also involves addition since we need to "undo" the subtraction in order to get back to the original amount of marbles. This certainly required her to do more than remember a fact or reproduce a skill. It required higher-order reasoning skills involving comprehension of the task, application of the concepts and appropriate skills, and analysis and some synthesis of the major concepts involved. Through further probing, this question allowed the teacher to see that MJ had little appreciation for the inverse relationship between addition and subtraction. She had learned to answer routine story problems without fully understanding the concepts.

Feature 2: Students Can Learn by Answering Good Questions, and the Teacher Learns about Each Student from the Attempt

Good questions have the potential to make students more aware of what they do know and what they do not know. That is, students can become aware of where their understanding is incomplete as well as where they have developed misunderstandings. Let's revisit the earlier question asked of MJ:

> *I have some marbles. I give some away to my friends and am left with fifteen. How many marbles might I have started with and how many might I have given away?*

The very act of trying to complete this question can help students gain a better understanding of the concepts involved. Imagine that MJ had told her teacher that one possible answer was 10, explaining, "It's 'subtraction' because the person is giving away or getting rid of marbles. I thought maybe the person gave away five. So I did fifteen minus five is equal to ten." By answering this question, MJ and her teacher both learn about her current level of understanding of how to use addition and subtraction to model story problems. MJ understands that the operation of subtraction can be applied to situations involving the "take away" or "removal" of quantities. And this is encouraging! However, in this problem, she has focused on the action without taking into account the larger context of the problem—the fact that 15 is what's left after some marbles were given away. With this revelation, MJ and her teacher can now work toward helping her reach a deeper level of understanding so that MJ can use the sequence of actions in a problem to choose an appropriate operation. Use of the good question not only reveals misconceptions to the teacher (proving to be a valuable assessment tool) but can empower MJ to use her mistakes as learning tools.

Feature 3: Good Questions May Have Several Acceptable Answers

Many of the questions teachers ask, especially during mathematics lessons, have only one correct answer. Such questions are acceptable, but there are many other questions that are open-ended, allowing multiple answers, and teachers should make a point of asking these, too. Each of the good questions that we have already looked at has several possible answers. Because of this, these questions foster higher-level thinking, encouraging students to develop their problem-solving expertise at the same time as they are acquiring mathematical skills. Good questions also dispel the myth that math problems always have just one right answer.

Feature 4: Good Questions Are Accessible to All Students

Good questions invite all students into the conversation. Because good questions allow for multiple entry points, students can use their prior knowledge as a gateway toward new understandings. Take, for example, the earlier question on place value:

> *I am thinking of a number between 10 and 100 with a single 9 in it. What might my number be?*

Some students may access this question using a model such as a number line or hundreds chart while others may rely on their fluency with the counting sequence or their facility thinking about two-digit numbers in terms of tens and ones. Because this question has multiple answers, it supports students with emerging, developing, or advanced understanding of place value. Good questions are accessible to all students because they do not require only one set pathway or approach.

Looking Ahead

In this chapter, we have looked more closely at the four main features that categorize good questions. The next chapter shares two ways to construct your own good questions.

CHAPTER 2

How Do I Create Good Questions?

When you first start using good questions, we suggest you select your questions from the collection of questions in Part II, "Good Questions to Use in Math Lessons." After a while, you will want to create good questions for yourself. Detailed in this chapter are two methods that can be used to create good questions. The one you use is a matter of personal preference.

Two Methods for Creating Good Questions

1. Working Backward
2. Modifying a Standard Question

When creating your own good questions, it's important to plan the questions in advance, because creating them is not something that can be done in the moment. Creating good questions relies heavily on the destination we have in mind. What do we expect our students to be able to do, say, or understand by the end of the lesson? Beginning with the end in mind requires us to start with a clear understanding of desired knowledge, learning, and outcomes. When we think about questions that we might ask our students, it is helpful to first ask ourselves a few questions; for example:

- What are the mathematical goals of the lesson?
- What are the connections we'd like students to make between lesson goals and previously covered concepts and/or procedures?
- What are the misconceptions students may have?
- What is our assessment of students' understanding?

Method 1: Working Backward

This method is a three-step process.

The Working Backward Method

Step 1. Identify a topic.
Step 2. Think of a closed question and write down the answer.
Step 3. Make up a question that includes (or addresses) the answer.

For example:

Step 1. The topic for tomorrow is averages.
Step 2. The closed question might be *The children in the Smith family are aged 3, 8, 9, 10, and 15. What is their average age?* (The answer is 9.)
Step 3. The good question could be *There are five children in a family. Their average age is 9. How old might the children be?*

Some more examples of how this method works are shown in the following table.

STEP 1 Identify a topic.	STEP 2 Think of a closed question and write down the answer.	STEP 3 Make up a question that includes the answer.
rounding	11.7	My coach said that I ran the race in about 12 seconds. What might the time on the stopwatch have been?
counting	4 chairs	I counted something in our room. There were exactly 4. What might I have counted?
area	20 in^2	Find as many different ways as you can to make a rectangle using exactly 20 square tiles.
fractions	$3\frac{1}{2}$	Two numbers are added to get $3\frac{1}{2}$. What might the numbers be?

(Continued)

STEP 1 Identify a topic.	STEP 2 Think of a closed question and write down the answer.	STEP 3 Make up a question that includes the answer.
money	35 cents	I bought some things at a supermarket and got 35 cents change. What did I buy and how much did each item cost?
graphing	x x x x x x x x x x x x x x x x x 1 2 3 4 5	What could this be the graph of?

Method 2: Modifying a Standard Question

This is also a three-step process.

The Modifying a Standard Question Method

Step 1. Identify a topic.
Step 2. Think of a standard question.
Step 3. Modify it to make a good question.

For example:

Step 1. The topic for tomorrow is measuring length using nonstandard units.

Step 2. A typical exercise might be *What is the length of your table measured in handspans?*

Step 3. The good question could be *Can you find an object that is three handspans long?*

Some more examples of how this method works are shown in the following table.

STEP 1 Identify a topic.	STEP 2 Think of a standard question.	STEP 3 Modify it to make a good question.
space	What is a square?	Write as many things as you can about this square.
addition	337 + 456 =	On a train trip I was working out some distances. I spilt some drink on my paper and some numbers disappeared. My paper looked like this: $\begin{array}{r} 3\,\square\,7 \\ +\,\square\,\square\,6 \\ \hline 7\,9\,\square \end{array}$ What might the missing numbers be?
subtraction	731 – 256 =	Arrange the digits so that the difference is between 100 and 200.
time	What is the time shown on this clock? 	What is your favorite time of day? Show it on a clock.

Looking Ahead

The more experience you have with good questions, the more you will want to use them, and the easier it will become for you to make up your own. Refer to either or both of the methods outlined in this chapter to help you feel confident in creating your own (in doing so, you might also "invent" additional methods!). In the next chapter, we outline five steps to successfully implementing good questions in your classroom.

CHAPTER 3

Five Steps to Successfully Implementing Good Questions in Your Classroom

Today's mathematics classrooms should be dynamic places where students are involved and engaged in their own learning. This can be achieved through activities that promote higher-level thinking, cooperative problem solving, and communication. We have seen that good questions support these activities and are readily available for teachers to use. But how can you ensure success with good questions? In this chapter we outline five steps we've found helpful in successfully implementing good questions, including advice for overcoming problems that might arise at each stage. We then take you through each of the steps using a specific good question.

Five Steps to Successfully Implementing Good Questions

1. Prepare the good question.
2. Present the good question.
3. Students work on the good question.
4. Facilitate a whole-class discussion.
5. Summarize.

Step 1: Prepare the Good Question

See Chapters 1 and 2 for guidance in understanding what makes up a good question as well as how to create good questions. Remember that

Part II of this resource offers many classroom-tested good questions for you to select and use as well.

Once you've selected a good question, it's critical that you understand the mathematics embedded in the question. On your own or with colleagues, think through the question and solve it. This is of particular importance with good questions because they might have more than one answer and/or approach. Try to think of more than just one answer. Doing so will help you anticipate and then react to the variety of responses you will hear from your students (undoubtedly, students will think of things you did not!). Remember, working through good mathematics takes time. An understanding of the mathematics involved in the question will help you determine the amount of time students might need to process and answer it.

Make sure the question is appropriate for your students; consider how students may begin the process of answering it. If needed, adjust the language and/or add more data to a question to ensure accessibility for all. Following are questions we've found helpful in working through a good question prior to teaching with it.

Questions for Preparing a Good Question

- Do I have the necessary materials (e.g., graph paper, chart paper, manipulatives)? If not, where can I get them?
- What misconceptions or difficulties might my students have with the language, concepts, or directions?
- What follow-up questions can I ask that will readdress or redirect misconceptions or difficulties?
- How much time will students need to answer the question?

Step 2: Present the Good Question

When presenting the good question to students, write it on the board where everyone can see it. As you ask the question, refer to the corresponding words on the board. Make sure that all students know what the question is; do not assume they know just because they can see it. Consider asking some students to repeat the question in their own words.

Allow time for students to ask you about the meaning of the question. Clarify the question to them if necessary but do not give any directions or suggestions on how to do it. This is for students to work out for themselves. The following is a summary of essentials we like to revisit before presenting good questions.

Essentials for Presenting Good Questions

- Present the question clearly using accessible mathematical language.
- Set clear and reasonable expectations for student work.
- Allow for individual approaches, methods, and/or answers.
- Ensure concrete materials are available for student use.

Step 3: Students Work on the Good Question

Once the question has been introduced, allow ample time for student investigation. We suggest having students work in pairs or small groups. This gives them the opportunity to communicate their ideas to others. Not only do students hear and learn from one another's answers, they also rehearse their thinking before sharing with the whole class (we've found that students are often more willing to share their reasoning and ideas with the whole class once they have first had a chance to practice with a small group). Encourage students to keep written records of their thinking together—calculations, charts, diagrams, and/or pictures. This documentation will help them support and defend their thinking and answers.

Pair or small-group work can also benefit students who may have difficulty starting. Students should be encouraged to go to each other first before seeking help from the teacher. Suggest that they try representing the problem in some way, such as by drawing a diagram or using materials (a variety of concrete materials should always be available for students to reference).

If, as students are working, you find that there are too many who cannot make progress without your assistance, you might need to reconvene as a class. Have students share their concerns. If the concerns of each group—or individuals within each group—are all different, then this is a sign that the question you have posed may need to be further modified. You could also decide to abandon the question altogether as unsuitable at this stage. If this happens, do not worry, because it takes time and practice to choose appropriate good questions. We assure you it's worth the effort and perseverance.

Once students are working, monitor their progress and note strategies, partial understandings, and misconceptions. If students stop after giving one response, ask them to look for other possible answers. If they have found all possible answers, ask them to explain their thinking behind each answer. You could also ask a related question to extend their learning. For example, you could change the parameters of the problem, add an additional step, or change the size of the numbers in the problem.

Step 4: Facilitate a Whole-Class Discussion

Facilitate a whole-class sharing of answers, strategies, and the discovered mathematics. This is a meaningful time for students to reflect and build on their learning. Class discussions offer opportunities for students to achieve understanding by processing information, applying reasoning, hearing ideas from others, and connecting new thinking to what they already know (Chapin, O'Connor, and Anderson 2013).

You do not have to wait until *all* groups have finished the task before initiating a discussion; it is better to stop while students are still engaged with the question and interested in the task. Consider signaling to groups that they have five minutes left before the whole-class discussion.

As students share, write their responses where everyone can see them and/or display their visual representation. If there are errors, work together to identify such and figure out why. Often, as students are explaining their work, they discover the error themselves.

Guide the conversation by keeping the focus on students' thinking. Encourage students to support, add to, and even disagree with one another's strategies. Have them address one another, not you. The following questions may help guide and facilitate discussions of students' answers.

Questions to Prompt Whole-Class Discussions

- Why do you think that?
- How did you know to try that strategy?
- How do you know you have an answer?
- Will this work with every number? Every similar situation?
- When will this strategy not work? What is a counterexample?
- Who has a different strategy?
- How is your answer similar to or different from another student's answer?
- Repeat your classmate's ideas in your own words.
- Do you agree or disagree with your classmate's idea? Why?

Step 5: Summarize

Typically, a summary of the main learning will naturally evolve as the whole-class discussion does. Continually check for understanding; just because a few students are eager to share does not mean everyone understands. To wrap up, review key points with everyone. Reference visual representations and use teaching aids as needed to support various learning styles. Relate the answers back to the question. It is also helpful to pose more questions using a similar format so that students can apply what they have learned to new situations.

The Five Steps in Action

Let's have a look at how the five steps might unfold in a third-grade classroom.

Step 1: Prepare the Good Question

Ms. Arnold wants a question to supplement the unit on whole-number addition that her students are learning. She selects the following question and solves it. With colleagues, she identifies possible misconceptions and difficulties she feels students may have including the idea that each number must be between 100 and 120 or confusion about the meaning of the term *consecutive even*. She notes to ask follow-up questions to redirect as needed: If you add your numbers, is the sum between 100 and 150? How are consecutive numbers similar to consecutive even numbers? How are they different? She notes, that if needed, she can make the question easier by changing the constraints of the problem to three consecutive numbers or two consecutive even numbers. Based on this preliminary work, she decides to allow approximately five minutes for students to investigate the question.

The Good Question

Three consecutive even numbers add up to a number between 100 and 120. What might the numbers be?

Step 2: Present the Good Question

Ms. Arnold writes the question on the board where all students can see it. She asks some students to read the question out loud and asks others to tell her what it means in their own words. She gives time for students to ask questions, being careful not to give any directions or suggestions on how to do the task.

Step 3: Students Work on the Good Question

Ms. Arnold has students work on the question in small groups. She circulates and observes. One group stops after discovering one response. She encourages the group to look for other possible answers. Several other groups find a few answers; she asks them to think of a way to describe all their answers—what do all of the answers have in common? For one group that finishes more quickly than the others, she poses a related task:

> *Four consecutive odd numbers add up to an even number between 120 and 150. What might the numbers be?*

When all groups have at least one response to the question, Ms. Arnold gives them a heads-up that they have five minutes before the whole-class discussion. She is OK with groups being at different stages in their work.

Step 4: Facilitate a Whole-Class Discussion

Ms. Arnold brings everyone together for a whole-class discussion. She asks students to share their answers. The answers from three different groups are:

- Group B: *There are many different answers.*
- Group E: *30, 40, 50*
- Group C: *We found two answers so far: 34, 36, 38 and 36, 38, 40*

Ms. Arnold notes that these three responses differ not only in the level of mathematical understanding but also in the quality of thinking that is demonstrated by the answers. She takes a positive approach to each group's response. She asks Group B to try to find one possibility. She asks Group E to talk about the meaning of the term *consecutive even*. She has Group C see if they can use their two answers to try to find additional possibilities. She continually encourages discussion using questions like:

- "Why do you think that?"
- "How is your answer similar to or different from another student's answer?"
- "Do you agree or disagree with your classmate's idea? Why?"

Step 5: Summarize

Ms. Arnold reviews the main points: consecutive even numbers must be both even and in order and a solution to the problem must satisfy *all* of the clues or given information. Even though these points were brought up in the discussion, she revisits them. She pushes students' thinking, first asking them to give examples of even numbers that are and are not consecutive and then to use mental math addition to check to see if students' suggestions are correct.

Ms. Arnold poses a similar task:

Three consecutive numbers add up to an even number between 180 and 200. What might the numbers be?

Looking Ahead

Now that you have an understanding of what makes a good question (Chapter 1), methods for creating good questions (Chapter 2), and the basics for using good questions in your classroom (Chapter 3), we invite you to explore the second part of this resource, which is filled with our favorite good questions. We hope that you too will find them useful as you incorporate more questioning techniques into your classroom.

Good Questions to Use in Math Lessons

CHAPTER 4	Good Questions for Money and Financial Mathematics	23
CHAPTER 5	Good Questions for Counting and Place Value	35
CHAPTER 6	Good Questions for Decimals	53
CHAPTER 7	Good Questions for Operations	65
CHAPTER 8	Good Questions for Fractions	81
CHAPTER 9	Good Questions for Geometry	93
CHAPTER 10	Good Questions for Data Analysis and Probability	115
CHAPTER 11	Good Questions for Measurement	135

In Part II, we offer examples of good questions for you to select from and use in your classroom. Each question is accompanied by notes to support your teaching.

The Organization of Questions

The questions are categorized by mathematical topic, each topic organized into two grade spans: grades K–2 and grades 3–5. For some topics such as decimals, there are questions only for grades 3–5.

Many of the questions in these spans can be adapted to meet the needs of the students in your classroom by making them easier or more difficult. You may wish to explore questions in both grade spans as you decide on appropriate questions for your particular class. For example, if you teach grade 2 and your students have a thorough understanding of whole-number addition and subtraction, you may want to explore questions in the grades 3–5 span within Chapter 7, "Good Questions for Operations."

In general, the questions in each span are neither hierarchical nor sequential. Each can be posed individually. Each span can be viewed as an à la carte menu of question choices to help students deepen their understanding of the mathematics they are studying.

The Investigations

As you are reading through the good questions, you will find some instances where they have been written as investigations rather than questions. This has been done where we felt they were better written as investigations. Use them in exactly the same way as the questions.

The List of Experiences

At the beginning of each topic for each grade span is a list of experiences that students should encounter for the particular topic. Not all students will be ready for these experiences at the same time. It is quite possible that some children in grades 3–5 might be working on some of the experiences listed for grades K–2. The list should not be treated as a progression of experiences but rather as a range of possible experiences.

The List of Materials

A list of materials that you might need is also provided at the beginning of each topic for each span. You will not need all of these materials unless you complete every question listed for the topic in that span. Check that you have suitable materials before you present a question to your class. It is important that students have a variety of concrete materials to select from when they are working on mathematical tasks.

The Teacher Notes

Following most questions there are teacher notes. Some identify important teaching points for the particular question while others describe in more depth the mathematics being addressed or misconceptions that may present themselves. At other times they will be useful in helping you assess students so you can plan to overcome any difficulties. It is a good idea to make notes as you observe students working to use in future planning.

CHAPTER 4

Good Questions for Money and Financial Mathematics

GRADES K–2	**24**
Money	*24*
GRADES 3–5	**30**
Money	*30*

Children's early work with money focuses on identifying coins, finding different ways to represent the same value (e.g., 25 cents can be made with 2 dimes and 1 nickel or 5 nickels) and using money as a context for making sense of various types of addition and subtraction problems. As children work with money in math class, they rely on their everyday knowledge and also begin to develop important foundational literacy skills. In later grades, children hone these skills by answering questions that require estimating values and prices as well as analyzing costs and savings.

Money (Grades K–2)

Experiences at This Level Will Help Children To

- recognize, describe, sort, and classify different coins
- exchange money for goods in play situations and give appropriate change
- order money amounts
- use coins to represent written money amounts and use numbers to record the value of a group of coins
- use estimation and a calculator for money calculations

Reproducibles are available in a downloadable, printable format. See page xx for directions about how to access them.

Materials

- coins and play bills
- goods marked with varying prices below $1.00 as part of the craft fair (Ensure that there are combinations of items that add to $1.00.)
- Cafeteria Price List (Reproducible 1)

Good Questions and Teacher Notes (pages 25–29)

1. Make 20 cents in as many ways as possible.

2. In my pocket I have 75 cents. What coins might I have?

In Questions 1 and 2, children should realize that there are many different ways to make a money amount. See if they use only multiples of one coin; for example, 4 nickels, as well as combinations of different coins, for example, 10 cents + 5 cents + 5 cents.

Are children confident when counting in fives, tens, twenties?

3. I bought something and got 5 cents change. How much did it cost and how much money did I give to pay for it?

Children's responses might be:

- costs 5 cents and gives 10 cents
- costs 15 cents and gives 20 cents
- costs 95 cents and gives $1.00
- costs $1.95 and gives $2.00

Can children see the folly of giving 15 cents for an item costing 10 cents to receive 5 cents change?

4. I spent exactly $1.00 at our craft fair. What might I have bought?

Check how children add amounts to $1.00. Note if they calculate multiples of 5, 10, or 20 to make $1.00; for example, do they know five items at 10 cents each is 50 cents or do they add each one separately?

5. I am a coin with a person on me. What might I be?

The main focus here is to look more closely at the attributes of coins.

6. I have two coins in one hand and one in the other hand. The coins in each hand are worth the same amount. What could the coins be?

Note if children develop a system when recording. How easily do they calculate amounts?

7. The answer to a calculation is 35 cents. What is the question? Refer to the following list to help you.

(See Reproducible 1, Cafeteria Price List.)

Cafeteria Price List

Peanut butter sandwich	$1.10	Salad	$1.55
Ham rollup	$1.40	Bag of chips	$0.65
Fruit salad	$1.15	Piece of fruit	$0.20
Cookie	$0.15		

Can children write more than one question?

8. I had one of each of the coins in our currency on my table. I sorted them into two groups. What might the groups have been?

It is interesting to note what categories children use. Ask them to tell you their categories; don't assume you know their reasoning.

9. The price tag on a keychain is $2.75. What coins would I use to pay for this?

Note if children develop a system when recording. How easily do they calculate amounts?

10. I have exactly $100 in bills in my pocket. What bills might I have?

Are children aware of available bills? Check if they can count in fives, tens, twenties, fifties.

11. Someone was asked to remember the cost of five items. They knew the most expensive was $2.00 and the least expensive was 50 cents. What might the other three be?

The focus here is on ordering of money amounts. Note if children can record different amounts correctly.

12. I bought an item at a shop and got 35 cents change. What did I buy and how much did it cost?

Children need to see the folly of including such things as buying an item costing 5 cents and giving 40 cents to get 35 cents change. Note if children look for a pattern when recording answers.

13. I gave change of $1.00 using quarters, dimes, and nickels. What might the change have looked like?

Note if children record systematically and accurately. Check how easily they make $1.00.

14. How could I spend exactly $20.00 at the supermarket? (Use a supermarket advertisement and a calculator to help.)

Check if children use estimation skills to help them; for example, they might round off some amounts to assist their estimation. Note how they use the calculator.

15. In my pocket I have $36.00. What bills might I have?

This allows you to see how familiar children are with the various bills and if they use a system when recording.

16. I bought something and paid for it with three coins. What might it have been and how much did it cost?

Look for a range of responses that are realistic.

Money (Grades 3–5)

Experiences at This Level Will Help Children To

- round to the nearest dollar to estimate or check total cost
- record money amounts
- pay with appropriate amounts when the exact amount is not available
- order money amounts
- use an appropriate method (mental, written, calculator) to solve problems involving money
- use mental calculation and estimation
- use symbols +, −, ×, and ÷ for written computation of money

Reproducibles are available in a downloadable, printable format. See page xx for directions about how to access them.

Materials

- coins and play bills
- supermarket advertisements from newspapers
- calculators
- Money Number Cloud (Reproducible 2)

Good Questions and Teacher Notes (pages 31–33)

1. I spent $80.00 on six tickets to the theater. How many adults and children are there and how much are the tickets?

Are the answers realistic? Can children multiply amounts, for example, 4 × $12.00 or 4 × $8.00? Note if they figure mentally or use paper and pencil to compute.

2. When I was in a shop I saw that a backpack cost about $22.00 and a lunch box about $15.00. What might have been the price tag on the backpack and the lunch box?

This question focuses on rounding off. Are children aware that they can round up and down?

3. A number sentence uses three of the following amounts or numbers in this cloud. What might the number sentence be?

(See Reproducible 2, Money Number Cloud.)

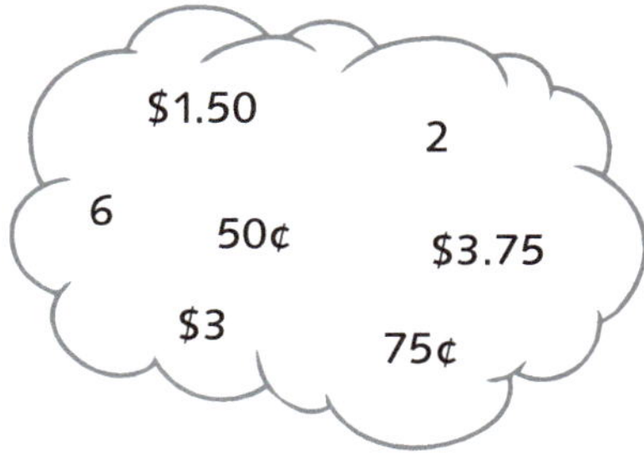

The main focus here is to see if children use a variety of processes, for example, 6 × $0.50 = $3, $1.50 ÷ 2 = $0.75, $3.75 – $0.75 = $3.00.

4. My friends and I shared an amount of money equally between us. We each got $1.20. How much money was there and how many friends might I have?

It is interesting to see how children do this—mentally, with paper and pencil, or with coins. When they check their answer do they include themselves or only the friends?

5. I went to get $100 out of the bank. What are the different ways I can ask for this amount of bills?

Note how children multiply and divide by 2, 5, 10, and 20.

6. Scientific calculators cost $40.00 and basic calculators cost $12.00. How much might it cost for a class set of some basic and some scientific calculators?

Note how children decide how many of each calculator to purchase. Do they record their answers systematically? Do they choose appropriate operations to work out the price? How easily do they handle these operations?

7. If one of the bills currently in use was to be changed to a coin, which one would you choose? Why?

Children should be able to justify their choice in a reasonable manner. You could extend this by looking at bills and coins in use in other countries.

8. Design a rounding policy for a supermarket.

The way children approach this will tell you if they understand rounding. It is interesting to note whose side they are on—owner or customer.

CHAPTER 5

Good Questions for Counting and Place Value

GRADES K–2	**36**
Counting	*36*
Place Value	*40*
GRADES 3–5	**44**
Counting	*44*
Place Value	*48*

When working with whole numbers, children must think about grouping by tens. For example, a group of ten ones is represented by one group of 10 and a group of ten tens becomes 100. One whole number can also be thought of in terms of different base ten units. For example, we can think of the number 300 as three groups of 100 as well as thirty tens or three hundred ones. If we take one unit and partition it into ten parts, each part represents one-tenth. If we take one-tenth and partition it into ten parts, each part represents one–one-hundredth. Just like whole numbers, decimal numbers can be understood in terms of different base ten units. For example, the number $\frac{3}{10}$ can be thought of as thirty-hundredths and the number 0.34 can be thought of as thirty-four–hundredths or three-tenths plus four-hundredths. Counting and counting methods become a central focus of children's work as they become facile with the cardinality rule and learn to count by a unit larger than 1. Children learn to compare and order numbers using concepts associated with place value and to work with numbers using the number line, hundreds chart, and base ten models. Finally, children learn that numbers belong to various classes and can describe a number as a member of one or more classes (e.g., 20 is both an even number and a multiple of 10).

Counting (Grades K–2)

Experiences at This Level Will Help Children To

- compare and order using one-to-one correspondence
- write, say, and count numbers to 10 and beyond
- skip-count forward and backward (ones, twos, fives, and tens to 100, and tens and hundreds to 1,000)
- use ordinal numbers up to 10
- place familiar two-digit numbers on a number line
- recognize odd and even numbers
- recognize patterns on a hundreds chart

Materials

- number charts
- number lines
- interlocking cubes

Good Questions and Teacher Notes (pages 37–39)

1. Write down everything you can about the number 12.

Not only can you see what children know, but they can become aware of what they themselves know. Repeat for other numbers.

2. Write down some odd numbers between 0 and 100.

Check that all numbers children write are odd. Do they record them haphazardly or methodically? See if children hold the misconception that an odd number is any number with an odd digit (e.g., 32).

3. I have written a secret number between 50 and 70. It is an even number. What might it be?

Can children write them all? Ask children to describe another characteristic of each number they write (e.g., 60 is even and it is also a number you say when you skip-count by tens). Repeat for other numbers.

4. Make a two-color train with interlocking cubes, using two of one color, two of the other, continuing for as long as you like. How many cubes are in your train?

Repeat for three colors. Or, ask children to build trains with three of each color.

5. I have written a secret number that is more than 65. What might it be?

Ask children to describe another characteristic of their number once they choose it (e.g., 99 is one less than 100). Extend this by asking children to write the biggest number they can. Compare these numbers to see which is the biggest. Repeat for other numbers.

6. Count by twos until you land on 20. What other numbers can you count by and still land on 20?

Repeat by increasing the goal number from 20 to 50, then 100.

7. Which number in this group does not belong? Why?

15 2 8 13 16

Children must explain their responses. Their answers can differ. Are their explanations coherent?

8. Starting at 0, what numbers can I skip-count by and land on 100?

Ask children to explain their results.

9. Show the following numbers on the given number lines. Write five other numbers on your number lines.

10 and 60

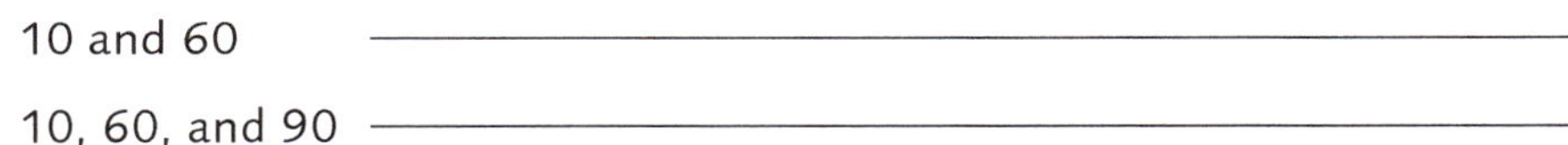

10, 60, and 90

Note if children realize that the largest number does not have to be put at the very end of the line. The placement of each number is dependent on the other number(s).

10. A number was shown as a set of dots. Part of the pattern looks like the following:

What might the number be? How do you know?

Accept any reasonable answers as long as children can justify them.

11. The number 8 is halfway between two numbers. What might those two numbers be?

How do children approach this one? Do they use a model such as a number line? The obvious answer is 7 and 9, but can children use this to derive other pairs of numbers such as 6 and 10, 5 and 11, 4 and 12 . . . ?

Place Value (Grades K–2)

Experiences at This Level Will Help Children To

- recognize, order, and write up to three-digit numbers
- develop early rounding ideas to 10
- recognize 0 as a number
- group items in tens to count larger collections
- use place value to work with patterns on a hundreds chart

Reproducibles are available in a download-able, print-able format. See page xx for directions about how to access them.

Materials

- base ten materials
- hundreds chart (Reproducible A)
- Jigsaw Piece (Reproducible 3)

Good Questions and Teacher Notes (pages 41–43)

1. What numbers can you make that are less than 100 and have 6 in the tens place?

This will indicate if children understand that the value of a digit depends on its position within a number.

2. I am thinking of a number between 10 and 100 with a single 9 in it. What might my number be?

Note if children are able to find all possible answers. Ask them how they can tell if they have them all.

3. You have a hundreds chart and jigsaw pieces. One of your pieces is shaped like the below. If that shape covers the number 43, what other numbers might it be covering?

(See Reproducible 3, Jigsaw Piece.)

Most children will initially need to refer to the hundreds chart, but encourage them to visualize as much as possible. This helps them to see the patterns that connect a particular number to its neighbors. There are twelve different answers to the question.

4. What do you know and what can you find out about the number 180?

Accept any suitable responses. Question further any children who do not include a response that shows their awareness of place value, for example, 180 is 100 + 80, or 18 groups of 10, and so on.

5. What numbers can you make using 6, 5, and 8?

Do children record their answers systematically and know when they have recorded all possibilities? Do any children include single digits—for example, 5—as one of their answers?

6. Find as many different ways as you can to make the number 40, adding only numbers 10 and 1. You may use each of these numbers as many times as you wish or not at all.

This question helps children to see that 40 can be named in many ways (e.g., two tens and twenty ones, three tens and ten ones).

7. Using base ten materials, show the number 25 in as many ways as possible.

After finding different ways, ask children which way uses the fewest number of blocks.

8. A two-digit number contains exactly one 4. What might the number be?

It is important to present the answers succinctly. How many different numbers might there be?

Counting (Grades 3–5)

Experiences at This Level Will Help Children To

- read and write whole numbers
- skip-count forward and backward by numbers from 2 to 10, starting from any number
- use materials to produce number sequences (e.g., square numbers)
- produce and describe number patterns
- recognize patterns in multiplication tables
- count beyond 1,000 starting from any number
- continue, create, and describe sequences involving constant multiplication and division or combinations of operations
- identify and work with prime numbers

Materials

- number charts
- calculators (useful for checking or producing number patterns)

Good Questions and Teacher Notes (pages 45–47)

1. What do you know and what can you find out about the multiples of 3 (3, 6, 9, 12, 15, 18, 21 . . .)?

One response could be that the sums of the digits of the multiples of 3 are multiples of 3.

2. Create a skip-counting pattern starting at 91 that someone else can continue.

It is important that children can describe their pattern to others. Have students share their patterns with other students and talk about how they created their pattern.

3. I doubled a number and kept doubling so that the original number was doubled four times. What might the answer be?

The answer will be a multiple of 16 (16, 32, 48 . . .). Do children use all the information in the question?

4. I halved a number and kept halving for a total of five times. The answer was a whole number. What number might I have started with?

It will be an even number and a multiple of 32. Ask children to describe how they found an answer.

5. My father is double my age. How old might I be?

6. My dog is half as old as me. How old might I be and how old is the dog?

7. My dog is half as old as me. My mother is double my age. How old might we each be?

For Questions 5, 6, and 7 the ages should be within a practical range. For example, in Question 7 ages of 2 (dog), 4 (me), and 8 (my mother); or 4 (dog), 8 (me), and 16 (my mother) are not acceptable. Note if children develop systems based on the inherent patterns to find responses.

8. Three consecutive even numbers add up to a number between 100 and 120. What might the numbers be?

Note how concisely children can communicate their answers. It is interesting to listen to them tell how they got their answers.

9. I am thinking of a number. The hundreds digit is larger than the ones digit. The tens digit is larger than the hundreds and it is odd. What might my number be?

Do children use all of the information to get an answer? Note if they check their answer against the question.

10. I am thinking of a number. If I divide by 3 there is a remainder of 1. If I divide by 4 there is a remainder of 1. What might my number be?

The pattern is key in this question. Are children able to describe it?

11. If I count by twos, I will land on both 100 and 1,000. If I count by threes, I won't land on either. What can you count by so that you don't land on 100 but you do land on 1,000? What can you count by to land on both 100 and 1,000?

Note the methods children use to investigate these patterns.

Place Value (Grades 3–5)

Experiences at This Level Will Help Children To

- read, write, compare, and order large whole numbers
- extend multiplication facts using place value (e.g., 3 × 5 = 15, so 3 × 5 tens = 15 tens)
- use place value to explain number patterns
- round numbers for estimation purposes

Reproducibles are available in a downloadable, printable format. See page xx for directions about how to access them.

Materials

- hundreds chart (Reproducible A)
- materials to model numbers, such as base ten materials

Good Questions and Teacher Notes (pages 49–51)

1. A number has been rounded to 1,200. What might the number be?

This depends on whether it has been rounded to the nearest ten or nearest 100. Numbers from 1,195 to 1,204 would round to 1,200 as the nearest ten, or numbers from 1,150 to 1,249 would round to 1,200 as the nearest hundred.

2. Write as many numbers as you can with 8 in the hundreds place.

Note if children only write numbers starting with 800 or if they write numbers more than 1,000.

3. Find as many ways as you can to rename 1,265 as the sum of smaller numbers.

How confident are children when moving between thousands, hundreds, tens, and ones? Two possible answers are:

a. 1,000 + 200 + 60 + 5
b. 1,000 + 100 + 150 + 15

4. Two numbers multiply to make 360. One of them has a 0 (zero) on the end. What might the two numbers be?

Can children find all possibilities? Can they see the pattern between pairs of numbers?

5. An easy way to add 9 is to add 10 and take away 1. Using a similar strategy what other numbers might I add or subtract in this way?

Children should be able to use other strategies such as to add 11 add 10 and then add 1 more; to subtract 99 take away 100 and add 1; and so on.

6. I wrote down a number with one 0 (zero) in it, but I cannot remember what it was. I know it was between 500 and 800. What might it have been?

Can children find all possible answers? How do they know they have found them all?

7. What numbers can you make using 1, 0, 2, 7, 8, and 4?

Children should record their answers methodically. You could ask them to write the largest or smallest number it is possible to make using all the digits.

8. Two numbers multiply to give 36,000. What might the two numbers be?

Yes, this is a place-value question! The key aspect is the zeros. Note how children handle them.

9. Write a number larger than one million. Write a number larger than ten million. How do you know the first number is larger than one million and the second larger than ten million?

Children should use place-value concepts to explain their choice of numbers.

10. Write a seven- or eight-digit number on the board. Ask children to write a number that is larger and a number that is smaller. Both must have the same number of digits as the number you wrote on the board.

Record many of the children's numbers on the board and discuss how they know each number is larger or smaller than the given number. Look for explanations that involve the value of the places in the numbers.

CHAPTER 6

Good Questions for Decimals

GRADES 3–5 54

Decimal Concepts 54

Our number system represents numbers by groupings of ten. When working with decimal numbers, children must think about partitioning into tens. For example, if we take one unit and partition it into ten parts, each part represents one-tenth. If we take one-tenth and partition it into ten parts, each part represents one–one-hundredth. Just like whole numbers, decimal numbers can be understood in terms of different base ten units. For example, the number $\frac{3}{10}$ can be thought of as thirty-hundredths and the number 0.34 can be thought of as thirty-four–hundredths or three-tenths plus four-hundredths. Because decimal numbers represent base ten numbers, children can extend their understanding of whole number addition and subtraction to add and subtract decimals (e.g., adding like parts only). When it comes to multiplication and division, however, children may be surprised by some of their "new" findings, including the discovery that the product may be smaller than either or both factors and the quotient may be larger than the divisor. With these two operations, it is especially important to connect children's work to meaningful interpretations (e.g., repeated subtraction) as well as concrete models of decimal numbers.

Decimal Concepts (Grades 3–5)

Experiences at This Level Will Help Children To

- read, write, order, and compare decimals to thousandths
- round a decimal number to the nearest whole number and use this skill to estimate
- rename decimals as common fractions and vice versa
- add, subtract, multiply, and divide decimals to hundredths

Reproducibles are available in a downloadable, printable format. See page xx for directions about how to access them.

Materials

- calculators
- materials to model decimals (e.g., base ten materials, Popsicle sticks, interlocking cubes, or metric rulers)
- six-sided or ten-sided dice
- strips of card or paper
- Decimal Number Cloud (Reproducible 4)

Good Questions and Teacher Notes (pages 55–63)

1. We are numbers that look like □ . □ and we are between 3.0 and 8.0. One of our digits is 6. What numbers might we be?

Do children record their answers systematically?

2. Use the digits 1, 2, 3, and 4 to create a true comparison:

Do children understand the function of the decimal point? Do they realize that the tenths digit on the left of the sentence does not have to be smaller than the tenths digit on the right of the sentence and that this is dependent on the size of the units digit? For example, 1.4 < 3.2 but 3.2 is not less than 1.4.

3. If I use a flat to represent one whole, a rod to represent tenths, and a unit to represent hundredths, what numbers can I represent using exactly ten pieces?

Children need to work with base ten blocks to do this question. You could extend it by asking them to show the biggest or smallest possible number using ten pieces.

4. A decimal number has been rounded off to 6. What might the number be?

This will establish if children understand the concept of rounding. Do they give one number or do they know it can include 5.5 but must be smaller than 6.5?

5. Players in turn roll a die (ideally, it can be a ten-sided die, but a six-sided die also works) and write the number in one of the squares:

□ . □ + □ . □ = 10

The aim is to make the answer to the addition as close as possible to 10. Once the number has been written, it cannot be erased or rewritten. This is as much a place-value question as a decimal question. The way children approach this will indicate if they understand the value of the places.

6. Show the numbers 0.1 and 1 on the given number line:

Write five other numbers on your number line.

Do children think that they have to place a 0 at the start of the number line or even that 0.1 has to be placed at the start? Do they think that 1 has to be placed at the end of the line? You could place 0.1 and 1 somewhere along the line and ask children to show the position of some numbers that could come before 0.1, between 0.1 and 1, and after 1.

7. My big sister says that the 100-yard dash record at her school is between 12 and 13 seconds. What might the record be?

In this question and the one that follows, note which children give answers using only hundredths or only tenths and which children give a combination of these. Are there any children who include thousandths? Does anyone give the complete range of tenths and hundredths?

8. In a race, the times are measured to hundredths of a second. The winner's time is 12.52 seconds. What might the times of the other eight runners be?

The main point is to check if children are able to order decimals. Note how realistic the times are. The times of the other runners must be greater than 12.52; for example, 12.53, 13.27, and so on.

9. Create a skip-counting pattern starting at 2.05 that someone else can continue.

Are children able to describe their pattern? How sophisticated are their responses? Let them check each other's patterns.

10. I wrote a sequence of numbers, adding the same number to each to get the next number. I wrote down 2.57 to start and 3.61 to finish. What might the numbers in between be?

Children could count by 0.02, 0.04, 0.08, or 0.26. Note which of them use a calculator to assist them to work this out. Do any children find the difference between 3.61 and 2.57 (1.04) and then find out which numbers divide evenly into it?

11. We are numbers that look like □ □ . □ and we are between 12.0 and 29.0. Our tenths digit is more than our ones digit. What numbers might we be?

Do children recognize the value of each place? Do they record their answers systematically?

12. Fill in the blanks to make a true comparison:

a. ______ -hundredths is less than ______ -tenths.

b. ______ -thousandths is greater than ______ -hundredths.

This question helps children differentiate between face value and place value and think about multiple representations of decimal numbers. For example, the number 0.34 (or $\frac{34}{100}$) is less than 0.8 (or $\frac{8}{10}$) because 0.34 has just three-tenths (and four more hundredths).

13. Fill in the blanks to make a true comparison:

a. ______ hundredths is less than $\frac{1}{4}$.

b. ______ hundredths is greater than $\frac{3}{4}$.

This question helps children connect decimals to common fractions. It can be adapted in many different ways (e.g., ______ thousandths is less than $\frac{1}{2}$).

14. Find decimal numbers to replace the missing numbers in the following inequality:

$$0 < \square < \frac{1}{3} < \square < \frac{2}{3} < 1$$

One possible strategy is to split $\frac{1}{3}$ in half. If children use this approach, watch for the common misconception that $\frac{1}{6} = 0.66666\ldots$

15. What might the missing numbers be?

$$\square . \square + \square . \square = 5$$

Do children realize that 5 is the same as 5.0? Check that they can give a range of answers.

16. Using only these keys on your calculator, what numbers can you make the calculator show?

The keys are: 5 . 4 + =

Children should use a range of processes. Note how comfortable they are with the functioning of the calculator.

17. I added three decimal numbers together to make exactly 4. What might the three numbers be?

Look for a variety of answers and strategies. For example, children can add two numbers and then subtract from 4 to find the third. Or they can find two decimal addends whose sum is 4 and then split one addend into two decimal addends.

18. One of your friends, Fred, asked you to help him with his addition calculations. He has done some calculations like this:

1.1 + 0.2 = 1.3

2.3 + 4 = 2.7

6 + 3.7 = 9.7

6.9 + 2 = 7.1

What are some other calculations that Fred might get wrong? What advice would you give Fred to help him?

The calculations that Fred might get wrong are those where a whole number is added to a decimal number and the decimal number is written first. Some other examples might be: 5.4 + 3 = 5.7, 2.8 + 6 = 3.4, and 7.5 + 4 = 7.9. The advice could include the fact that whole numbers have a decimal point after them but that for convenience we do not write it. For example, 4 is the same as 4.0.

19. In this calculation some numbers are missing. What might they be?

$$\begin{array}{r} 3.\,\square\,\square \\ +\,\square.\,7\,\square \\ \hline 6.\,\square\,3 \end{array}$$

Do children give more than one answer and do they record their answers systematically? Do they realize that ten different digits can be inserted on the top line?

20. I am thinking of two numbers. Neither number is a whole number. Their difference is 3.1. What might be the numbers?

Check to see if children use the inverse relationship between addition and subtraction to answer this question. For example, we can add 3.1 to any decimal number to find two possible numbers (e.g., 3.1 + 2.5 = 5.6 so the two numbers could be 5.6 and 2.5).

21. Is it possible to subtract two decimal numbers and get a difference that is a whole number?

Use a number line to help children generalize about all possible answers to this question (i.e., pairs of decimal numbers with a whole number difference).

22. Two numbers multiply together to give a product between 14 and 15. What might these numbers be?

It is useful if children have calculators to help with this one. For example, they might divide a decimal such as 14.4 by various numbers to find a solution (e.g., $14.4 \div 2.5 = 5.76$, so $5.76 \times 2.5 = 14.4$).

23. I multiplied 15 by a decimal and got an answer less than 15. What might be the decimal that I multiplied 15 by? Give at least ten possibilities.

Can children generalize that the decimal fraction has to be less than 1?

24. The area of a rectangle is 0.9 square units. What might the perimeter be?

By using a calculator, children can discover many possible answers (e.g., $0.9 \div 0.25 = 3.6$, so two sides of the rectangle could be 0.25 and 3.6, which means the perimeter would be 7.7). You could ask the children to find the largest or smallest perimeter to encourage them to find a pattern.

25. Find as many different ways as you can to make your calculator show a number with a particular decimal such as 12.34 without pressing the decimal point button.

Some possible answers are 1234 ÷ 100; 1234 ÷ 10 ÷ 10; 2468 ÷ 200; 617 ÷ 50; 1234 × 1 percent (if available).

26. I divided 6.12 by 3 and wrote down the answer, 2.4. What did I do wrong and what other similar calculations might I get wrong?

It is a common mistake for children to leave out the 0 in such examples. If they are able to give other examples then they are aware of the error.

27. Make as many different equations as you can using the numbers in this cloud.

(See Reproducible 4, Decimal Number Cloud.)

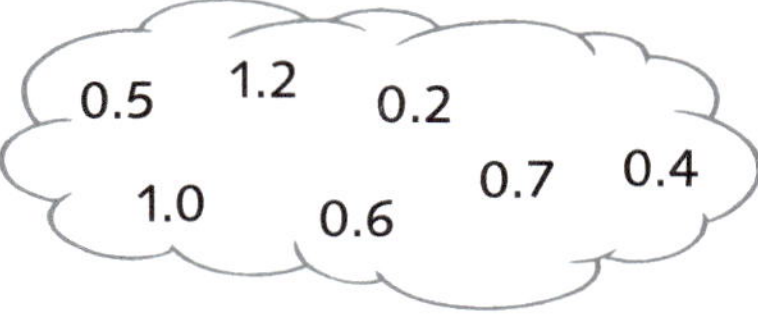

Look for a variety of operations.

CHAPTER 7

Good Questions for Operations

GRADES K–2 66

Addition and Subtraction 66

GRADES 3–5 72

Addition and Subtraction, Multiplication and Division 72

The four operations (addition, subtraction, multiplication, and division) are among the most important topics within children's elementary mathematics experience. In the younger elementary grades, children learn to extend their counting strategies to develop strategies for adding and subtracting numbers. They also begin to think of numbers in terms of groups (e.g., 12 can be thought of as two groups of 6). This type of thinking provides the foundation for formal work with whole-number multiplication and division later on. Children also learn that each operation is used to represent and solve many different types of problems. They develop meaningful computational strategies for each operation while also discovering how the four operations relate to each other—especially the idea that addition and subtraction and multiplication and division are inverses, respectively. Finally, the mathematical properties associated with each operation (e.g., the commutative property of addition) prepare children for later work with fraction and integer operations.

Addition and Subtraction (Grades K–2)

Experiences at This Level Will Help Children To

- use materials to explore how numbers can be decomposed
- learn about strategies for addition (e.g., using doubles, counting on, etc.)
- add and subtract small numbers in story problems
- write number sentences and make up stories about number sentences

Materials

- counters
- toy vehicles (with different numbers of wheels)

Good Questions and Teacher Notes (pages 67–71)

1. A basketball player scored 9 points in two games. What might her scores in each of the games be?

2. In the following problem, what might the missing numbers be?

   □ + □ + □ = 13

The purpose of Questions 1 and 2 is for children to think about the addition process in a different way and to show that there can be a range of possible answers to a given problem.

3. There are now four chickens in Mrs. Farmer's pen. How many chickens did she once have, and what happened to them?

Note the children who use larger numbers confidently. It is important to use counters or similar aids.

4. Make up some different ways to add 5 to 8 in your head. In how many ways can you do it?

This is to show that there is not only one correct way to do a calculation. You can extend this question by changing the size of the addends to include one or more two-digit numbers.

5. The difference between two numbers is 5. What might the two numbers be?

This focuses on difference as subtraction. Some children will feel confident using larger numbers. Some may record their answers systematically, for example, 6 – 1, 7 – 2, 8 – 3, 9 – 4, and so on.

6. Choose a number between 5 and 20. Write all the subtractions you can using that number. For example, if the number is 11 you could write:

11 – 6 = 5

15 – 11 = 4

13 – 2 = 11

Note those children who can generate strings of related facts and use the number in different positions in their subtractions, that is, the starting number, the amount taken away, or the answer.

7. Make up a sentence with five words but no more than twenty letters.

At this stage children will most likely use the total of 20 to work this out rather than balancing the number of letters above and below 4. In finding a response, children will use both addition and subtraction.

8. Yesterday I put some counters into groups with the same number in each group. I cannot remember the groups, but I can remember that there were twelve counters. What might the groups have been?

Children need counters to do this. The possibilities are two groups of 6, three groups of 4, four groups of 3, or six groups of 2.

9. There are five vehicles in the parking lot. How many wheels might there be?

Let children look at toy vehicles with different numbers of wheels. Do children add numbers of wheels separately or do they use doubles or other suitable strategies? You could look at situations where the number of wheels might be an odd number.

10. When the children in a class each got a partner, there was one child left over. How many children might there be in the class?

This would be good to model with the class.

11. I subtracted an odd number from an even number and got the answer of 41. What might the odd and even numbers be?

Do children record their number sentences systematically? Can they describe the pattern they find? This task could lead to an investigation of the result of adding or subtracting two odd numbers or two even numbers or one odd number and one even number.

12. What might the missing numbers be here?

$$\begin{array}{r} 3\,\square \\ +\;1\,\square \\ \hline \square\,2 \end{array}$$

This is to show that there are many possible answers. You could ask children to find all possibilities. Note those children who have difficulty working out solutions with trading.

13. Last night I added together two numbers, each with two digits. I got an answer of 67 but I cannot remember what the numbers were. Help me work out some possibilities.

Do children approach this systematically? The question is similar to the previous one but is stated in a more abstract manner.

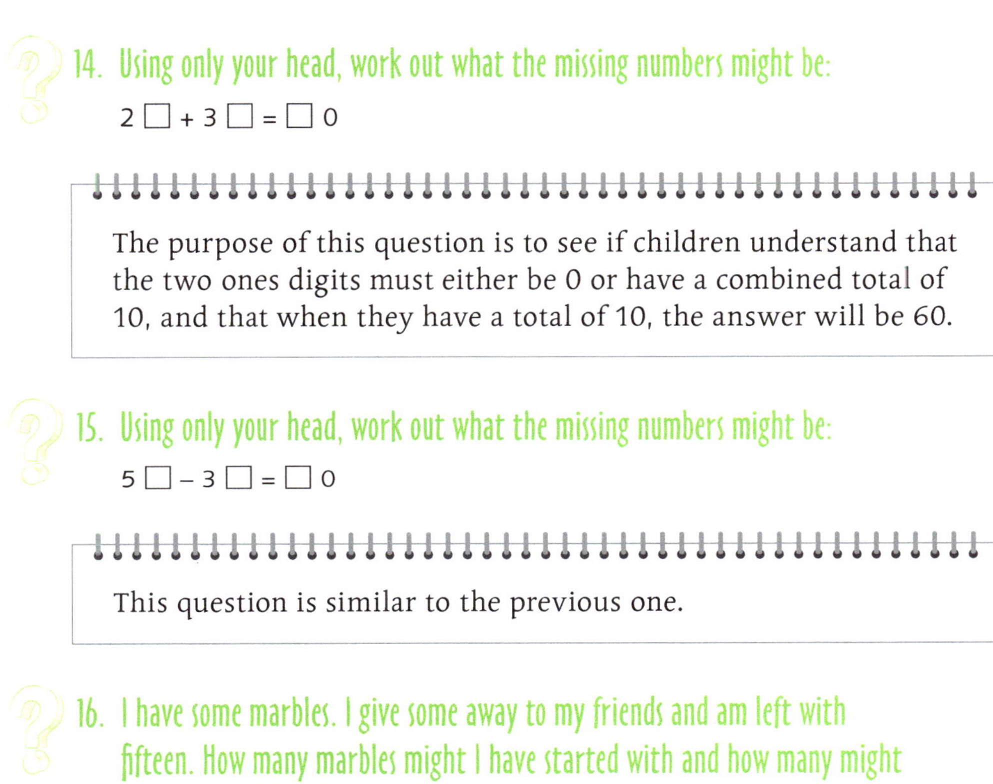

14. Using only your head, work out what the missing numbers might be:

2 □ + 3 □ = □ 0

The purpose of this question is to see if children understand that the two ones digits must either be 0 or have a combined total of 10, and that when they have a total of 10, the answer will be 60.

15. Using only your head, work out what the missing numbers might be:

5 □ – 3 □ = □ 0

This question is similar to the previous one.

16. I have some marbles. I give some away to my friends and am left with fifteen. How many marbles might I have started with and how many might I have given away?

Again, look for a range of answers. Note the size of the numbers children are confident working with.

Addition and Subtraction, Multiplication and Division (Grades 3–5)

Experiences at This Level Will Help Children To

- build on known number facts and use extended number facts
- double and halve
- develop mental strategies and estimation skills
- refine methods for addition and subtraction of whole numbers and decimals
- record simple multiplication and division calculations
- select the appropriate operations to solve number problems
- recall multiplication and division facts up to and including 10 × 10
- choose and use mental strategies
- estimate and check reasonableness of answers
- use suitable written methods for multiplication and division

Materials

- calculators
- concrete materials where necessary (e.g., counters, base ten materials, etc.)
- numbered cube

Good Questions and Teacher Notes (pages 73–79)

1. Five numbers added together make an odd number. What do you know about the numbers?

Either one, three, or five of the numbers must be odd. This question highlights some features of odd and even numbers.

2. The faces of this cube are numbered consecutively. What might the sum of the faces be?

The faces we cannot see could be numbered 10, 11, and 12 or 4, 5, and 6, or in between, so the sum could be 39, 45, 51, or 57. Let children use a calculator to add the numbers. You will need to discuss what *consecutively* means.

3. Eighteen people said they wanted to do folk dances. The teacher said they must dance in groups, but no one must be left out. Find as many different ways as you can to make different groups of dancers.

This is the same principle as Question 8 in the previous section. Let children use counters to represent the dancers.

4. I did a subtraction task and the answer was 215 but I cannot remember the other numbers. Find as many solutions to this subtraction as possible.

Which children develop a system? Note the numbers children use to do this task for they will tell you a lot about the stages they are at. For example, do they record subtractions like 692 – 477? When children are working out the number do they use addition to find the result and then use the number and the result to record a subtraction?

5. A number is divided by 5 and leaves a remainder of 3. What might the number be?

Do children do this by trial and error or do they develop a rule (i.e., multiply any number by 5 and add 3)?

6. Your friend Mira is trying to work out how to double numbers in her head. When she was asked to double 34 she said 68. When she was asked to double 23 she said 46. When she was asked to double 27 she said 44. When she was asked to double 36 she said 62. What are some other double calculations that Mira might get wrong? What might be your advice to your friend?

The advice could relate to the exchanging of ten ones for one ten when the ones digit that is doubled results in an answer of 10 or more, and adding it to the other tens.

7. What might the missing numbers be in this problem?

□□ ÷ □ = 3

There are 6 possibilities: 12 ÷ 4, 15 ÷ 5, 18 ÷ 6, 21 ÷ 7, 24 ÷ 8, and 27 ÷ 9. Do children use their knowledge of the 3 times table to help them?

8. The answer to a division calculation is 5. What might the calculation be?

Note how children do this. Do they know that if they multiply 5 by any number they will find the numbers to make their calculation, for example, 5 × 20 = 100, so 100 ÷ 20 = 5?

9. Work out all possible answers for this addition problem:

2□6 + □8 = □2□

There are nine possible answers and ten if you allow a 0 to be used before the 8.

10. Using the following digits and any operations, what numbers can you make?

9, 8, 7, 6, 5, 4, 3, 2, 1

Children could also be asked to find the largest or smallest number they can make.

11. What might the missing numbers be here?

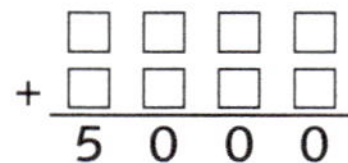

Do children realize that there are many possible solutions? Can they find the range of possible solutions?

12. I did a subtraction problem last night but can only remember the answer and that it looked like this:

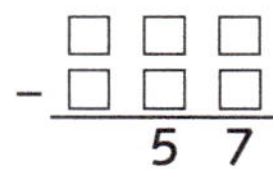

What might the missing numbers be? Find several different answers.

This question focuses on the difference between numbers. The smallest possible top number is 157 – 100 = 57 and the largest is 999 – 942 = 57. There are 843 possible combinations.

13. MT and JR each have a large collection of trading cards. JR has six times as many cards as MT. How many cards might they each have?

While most textbooks tend to focus on equal-groupings multiplication problems, children need experience with multiplicative compare problems as well.

14. A school has 400 students. They all come to school by bus and each bus carries the same number of students. How many students might there be on each bus?

Note the methods children use to do this. Do they use division or multiplication?

15. What might the missing numbers be?

□ 0 ÷ □ = □

There are 8 possibilities:

10 ÷ 2 = 5	30 ÷ 5 = 6
10 ÷ 5 = 2	30 ÷ 6 = 5
20 ÷ 4 = 5	40 ÷ 5 = 8
20 ÷ 5 = 4	40 ÷ 8 = 5

16. Solve these problems.

a. Using 9, 8, and 7 written as □ □ × □ , find as many different answers as you can.

b. Using 9, 8, 7 and 0 written as □ □ × □ □ , find as many different answers as you can.

c. Using 9, 8, 7 and 0 written as □ □ □ × □ , find as many different answers as you can.

Note the children who do this systematically.

17. In the following problem, what might the missing numbers be? Find as many different answers as possible.

$\square \times \square = 2{,}280$

Again, which children do this by division and which do it by trial-and-error multiplication? Do they use estimation skills?

18. How could you calculate 23×4 if the 4 key on your calculator is broken?

For example, children could press 23 + = = = = or 23 + 23 + 23 + 23 or 23 × 2 × 2.

19. Eighty-four children in four grades are arranged into teams with the same number on each team. How many teams are there and how many children might there be on each team?

Note the methods children use to do this. Do they use division or multiplication? Do they use known number facts?

20. Write a word problem where the answer is $27\frac{1}{4}$. Write a word problem where the answer is 27 and 1 remainder.

Are children able to deal with remainders in a sensible manner and know when to convert them to fractions and when to leave them as remainders? For example, if you share 109 pizzas among four families each family would get 27 whole pizzas and the remaining one could be cut into quarters. If you are putting 109 people on four buses you would put 27 people on each bus and fit the remaining person on one of the buses.

21. What could you add to 361 to make it divisible by 10?

Adding 9 is the obvious answer, but there are many other possible solutions. Can children find a rule for this one?

22. Using four 4s and any operation, find as many different answers as you can make.

It is possible to make all numbers between 1 and 100!

CHAPTER 8

Good Questions for Fractions

GRADES K–2 82

Fraction Concepts 82

GRADES 3–5 86

Fraction Models, Comparing Fractions, Adding and Subtracting Fractions 86

Elementary children need to develop deep understanding of fraction concepts and competence with fraction skills in order to be prepared for the study of algebra in the middle grades. Classroom experiences must move beyond fractions as parts of a whole to include other interpretations of fractions, including fractions as parts of a set and fractions as locations on a number line. The unit fraction becomes a focal point of children's work as they begin to think about any fraction $\frac{a}{b}$ as iterations of the related unit fraction $\frac{1}{b}$ (where b is not equal to 0). Children rely on their experiences with fraction models to order and compare fractions and form important generalizations about equivalent fractions. All of these concepts provide the foundation for work with fraction operations.

Fraction Concepts (Grades K–2)

Experiences at This Level Will Help Children To

- use informal fraction language for objects and collections
- compare fractional parts of objects and collections

Materials

- fraction materials such as rods, counters, or shapes
- lengths of string, tape, or yarn

Good Questions and Teacher Notes (pages 83–85)

1. You see a sign in a shop window that reads $\frac{1}{2}$ OFF SALE. What does this mean to you?

Listen carefully to children's responses as these will indicate the depth of their understanding. Also, give them some prices, ask them how much the sale price would be, and ask them to explain their reasoning.

2. Half of the people in a family are males. What might a drawing of the family look like?

Do children understand that there must be the same number of people in each group? Make sure they see a range of drawings done by class members.

3. Draw a shape. Show how to cut the shape into two halves.

Do children show equal parts?

4. I was listening to the radio and I heard the announcer say, "half." What might she have been referring to?

The answers will indicate depth of understanding. Two possible responses are "Half past three" and "Halftime."

5. We want to paint the top half of the room. How could we find out where the halfway mark is?

Allow children to do this how they want but have some string, tape, and so on, available for their use. The main focus here is to look at how practical the children's strategies are.

6. What do you know and what can you find out about $\frac{1}{4}$? Record it on paper or show it with materials.

Note if children understand $\frac{1}{4}$ as part of a whole and as part of a collection. Can they use a range of materials to represent it?

7. Make as many different designs as you can that are $\frac{3}{4}$ red and $\frac{1}{4}$ yellow.

Note if the designs are simple or complex. Ask children to explain how they know $\frac{3}{4}$ is red and $\frac{1}{4}$ is yellow.

8. One-third of a class orders lunches from the cafeteria each day. How many students might be in the class and how many of them order lunches each day?

Let children use counters to represent the students if they wish. Can they find more than one answer? Do they base their answers on multiples of 3?

9. My aunt said that when she was half her age she could touch her toes. How old might she be now and how old was she when she could touch her toes?

Check that suggested answers are realistic.

10. I picked up a handful of M&M's. One-third of them were red. What might a drawing of the M&M's look like?

This requires children to show a fractional part of a collection. Do children provide a range of answers and does anyone develop a system to do so? Do children understand why amounts that are not multiples of 3 do not work?

Fraction Models, Comparing Fractions, Adding and Subtracting Fractions (Grades 3–5)

Experiences at This Level Will Help Children To

- represent simple fractional parts of objects and collections
- order and compare fractions with the same denominators
- write common fractions
- record simple equivalence (e.g., $\frac{1}{2} = \frac{2}{4} = \frac{3}{6}$)
- add and subtract tenths and fractions with like or related denominators
- use equivalence to compare and order fractions
- locate fractions on a number line
- rename fractions in different forms (e.g., as percentages or decimals)
- mentally add and subtract common equivalent fractions
- understand the relationship between division and fractions

Materials

- fraction materials (e.g., counters, shapes, rods, or kits)
- drawing paper for designs, such as origami or other square paper
- circles cut into quarters to represent "pizzas"

Good Questions and Teacher Notes (pages 87–92)

1. My friend and I ate all of a pizza which was cut into 8 equal pieces. What fraction of the pizza might each of us have eaten?

Note children who work this out systematically.

2. What might the missing numbers be?

$\frac{\square}{5} + \frac{\square}{5} = \frac{\square}{5}$

Do children realize there is more than one possible answer?

3. I had some pizzas that I cut into quarters. How many pizzas might I have had, and how many quarters might I have after cutting them?

Can children identify a relationship between wholes and quarters?

4. What might the missing numbers be?

$\frac{1}{\square} = \frac{2}{\square} = \frac{3}{\square}$

Do children develop a pattern when answering this? Can they continue the pattern? If 10 was the numerator for their pattern, what would be the denominator?

5. Show $\frac{2}{3}$ as many different ways as you can.

Note if children understand $\frac{2}{3}$ as part of a whole or part of a length and as part of a collection. Do they use a range of materials to represent it?

6. I folded an origami square to show a fraction. How did I fold it and what might the fraction have been?

Look for equal parts and a range of answers.

7. Someone was counting by fractions and the last thing they said was, "ten." What might the four numbers before this have been?

How easily can children choose fractions before 10 in order, e.g., 10, $9\frac{3}{4}$, $9\frac{2}{4}$, $9\frac{1}{4}$, 9 or 10, $9\frac{9}{10}$, $9\frac{8}{10}$, $9\frac{7}{10}$, $9\frac{6}{10}$? Do they provide a range of answers?

8. A friend of mine put the following fractions into two groups. What might the two groups be?

$\frac{3}{4}, \frac{2}{5}, \frac{1}{3}, \frac{6}{10}, \frac{1}{10}$

Ask children to give reasons for their groups, as these could highlight some misconceptions. One possible grouping is to put $\frac{1}{3}$ and $\frac{1}{10}$ in one group because they are unit fractions; another grouping is to put $\frac{3}{4}$ and $\frac{6}{10}$ in one group because they are greater than $\frac{1}{2}$.

9. Two fractions add up to $\frac{1}{2}$. What might those two fractions be?

Do children only use known fraction combinations such as $\frac{1}{4} + \frac{1}{4}$ or do they use subtraction to find other possibilities; for example, $\frac{1}{2} - \frac{1}{3} = \frac{1}{6}$, so $\frac{1}{3} + \frac{1}{6} = \frac{1}{2}$? Do they use equivalence; for example, $\frac{1}{2} = \frac{6}{12}$, so $\frac{1}{12} + \frac{5}{12} = \frac{1}{2}$?

10. Some numbers add up to 10. I know that at least one of them has a fraction part in it, but none uses decimals. What might the numbers be?

Note the methods children use to find answers. These will tell you a lot about their understanding of fractions.

11. What three fractions might I add together to get an answer of $\frac{1}{2}$?

Again, look at the methods used. Do children guess and then work it out to check? If so, how do they then adjust the fractions? Do they know which fractions are smaller than $\frac{1}{2}$?

12. The answer is $\frac{3}{7}$. What might the question be?

Encourage children to use other processes than just addition.

13. What two fractions might I subtract to get an answer of $\frac{3}{4}$?

Do children use equivalence ("I know $\frac{3}{4} = \frac{6}{8}$, so $\frac{7}{8} - \frac{1}{8} = \frac{3}{4}$") or some other method to do this?

14. What might the missing numbers be?

$\frac{1}{\square} \times 3\square = 1\square$

The missing numbers do not have to be the same. Can children describe all of the answers? Can they prove they have all of the answers? (There are nine possibilities.)

15. Is the following student reasoning always correct? If not, give examples where it does not work.

Teacher: Tell me a fraction between $\frac{1}{2}$ and $\frac{3}{4}$.

Student: Two-thirds, because 2 is between 1 and 3, and 3 is between 2 and 4.

This question highlights a misconception that some children may have.

16. A rectangle has a perimeter of two units. What might the area be?

Yes, this is a fraction question! The perimeter must be a combination of fractions, for example, $\frac{1}{2} + \frac{1}{2} + \frac{1}{2} + \frac{1}{2}$, $\frac{1}{4} + \frac{3}{4} + \frac{1}{4} + \frac{3}{4}$, and so on. The area will vary depending on the length of the sides. You may want to remind children that a square is a rectangle.

17. Write some different stories about $3 \div \frac{1}{2}$.

The purpose of this is for children to understand the difference between $3 \div \frac{1}{2}$ and $\frac{1}{2}$ of 3. Their stories will indicate this. Stories like "I had $3.00 and I gave half to my friend" are not appropriate.

18. Listen to or read the following conversation between a teacher and a student.

Teacher: Which is bigger, $\frac{201}{301}$ or $\frac{2}{3}$?

Student: Well, $\frac{201}{301}$ is bigger because 1 has been added to the top and the bottom.

Is this reasoning correct? Are there any examples where adding 1 to the top and the bottom makes the fraction bigger?

This question highlights a misconception that some children may have.

19. What might the missing fraction be?

$\frac{\square}{\square} < \frac{3}{4}$.

Provide concrete materials for this question. Do not assume that because some children write, for example, $\frac{1}{3}$, that their understanding is correct. Check why they write this. Some children may think any number smaller than the 3 or the 4 will make a smaller fraction and will not consider fractions such as $\frac{1}{5}$, $\frac{5}{10}$, and so on, to be smaller. This question checks the same misconception as question 18.

20. What might the missing numbers be?

$\frac{\square}{\square} + \frac{\square}{\square} = \frac{2}{5}$

Do children use fractions other than those with a denominator of 5?

21. What might the missing numbers be?

$1\frac{\square}{\square} = \frac{\square}{\square}$

Can children draw a link between mixed numbers and other fractions? Can they describe how they found an answer?

22. Write down ten fractions between $\frac{1}{3}$ and $\frac{2}{3}$.

This question can also be posed as:

$\frac{1}{3} < \frac{\square}{\square} < \frac{2}{3}$ and also as $\frac{1}{3} < \frac{\square}{\square} < \frac{\square}{\square} < \frac{\square}{\square} < \frac{2}{3}$

CHAPTER 9

Good Questions for Geometry

GRADES K–2	**94**
Two-Dimensional Shapes	*94*
Three-Dimensional Shapes	*98*
GRADES 3–5	**104**
Two-Dimensional Shapes	*104*
Three-Dimensional Shapes	*111*

The two topics included in this chapter are:

1. two-dimensional shapes
2. three-dimensional shapes

Because we use spatial knowledge for a wide range of practical tasks, such as finding our way around, designing, and so on, it follows that many of the experiences in this chapter can be done in a practical manner using concrete materials.

Children can be encouraged to handle and make things using a variety of materials, to produce models, and to change the size, shape, and position of objects. They should be given many opportunities to talk about their work. Their spatial language will develop as they describe their own experiences and listen to others describe theirs.

In the upper-elementary grades, children need experiences that allow them to look more closely at the properties of shape and design and thus refine their thinking and descriptions.

The children's experiences when working on these questions will assist them to develop their spatial concepts and allow you to build up an accurate picture of each child's knowledge.

Two-Dimensional Shapes

(Grades K–2)

Experiences at This Level Will Help Children To

- use and understand language of relative positions in space, such as under, behind, up, between, and so on
- make and draw reasonable representations of common shapes (e.g., triangles)
- recognize and name common shapes
- match two-dimensional figures to faces of three-dimensional shapes
- use appropriate language to talk about shapes (e.g., round, corner, side, etc.)
- make pictures and patterns with shapes
- recognize symmetry and make symmetrical pictures

Reproducibles are available in a downloadable, printable format. See page xx for directions about how to access them.

Materials

- pattern blocks
- Triangles (Reproducible 5)
- Mixed Two-Dimensional Shapes (Reproducible 6)

Good Questions and Teacher Notes (pages 95–97)

1. Where could you stand in this room so the door is to your left/right?

This will let you see if children know left and right. Do it for other objects in the room too.

2. What is it that makes these shapes triangles?

(See Reproducible 5, Triangles.)

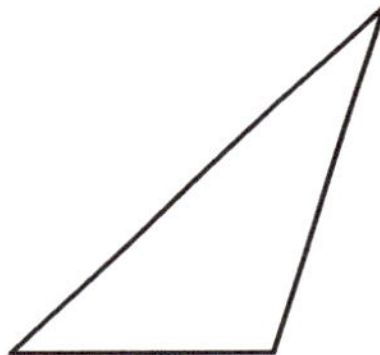

This makes children think about the properties of triangles rather than just draw any shape with three sides. It allows you to see if they are aware of any other properties of triangles other than three sides.

3. I drew a shape with four sides. Draw what my shape might look like.

It is a good idea to have a shape drawn (but hidden) so children can compare their shapes with it. It is important that children see all the four-sided shapes that are produced. One way to extend this question is to add the constraint that none of the four sides are the same length.

4. Peter said that he used two smaller shapes both the same to cover his large shape. If his large shape looked like this what might the two smaller shapes look like?

Give children paper rectangles, all the same as the sample, so they can experiment cutting it into two pieces in different ways.

5. A friend of mine sorted these shapes into two groups. What might the two groups have been?

(See Reproducible 6, Mixed Two-Dimensional Shapes.)

This allows children to group shapes according to properties. For example, they could choose regular/irregular shapes, curved/linear shapes, four-sided/others, and so on. Ask them to tell you why they have chosen the groups; do not assume you know their reasoning.

6. I made a picture using only circles and squares. What might my picture have looked like?

Check to see if children do only use circles and squares. Display finished pictures.

7. Make a design with pattern blocks. What did you make?

Children will explore how shapes can fit together.

8. Write down everything you know and everything you can find out about this square.

The main purpose of this question is for children to become aware of what they know. They should also see that even in simple figures there are many mathematical concepts. Their responses can include comments on: equal sides; equal angles; actual measurements of sides; the perimeter; the area; the lengths of the diagonals; the symmetry; and so on.

Three-Dimensional Shapes (Grades K–2)

Experiences at This Level Will Help Children To

- use and understand language of relative positions in space, such as under, behind, up, between, and so on
- follow directions and give oral directions
- build from imagination, memory, visual instruction, or oral description
- choose pieces that meet functional requirements
- copy simple arrangements of shapes
- identify and name common three-dimensional shapes in the environment
- use and understand functional spatial language (e.g., stacks, rolls)
- use simple language to describe objects
- classify shapes according to function and attributes
- distinguish between a three-dimensional object and its faces
- investigate the shape of cross sections of three-dimensional objects

Reproducibles are available in a downloadable, printable format. See page xx for directions about how to access them.

Materials

- lots of three-dimensional objects (e.g., cylinders, boxes, cones)
- a cardboard box
- blocks and other stacking shapes
- Mixed Three-Dimensional Shapes (Reproducible 7)

Good Questions and Teacher Notes (pages 99–103)

1. I can see something above the bookshelves and below the blinds. What could it be?

Adapt this to suit your situation. You could turn it into an *I Spy* game. The purpose is to assess if children understand the meaning of location language.

2. In my hand I have an object that is able to roll. What might it be?

Children should realize that a ball is not the only object that will roll. You could provide a variety of objects for them to try.

3. What objects in this room could pass through this opening?

Make a rectangular opening about 2 inches by 8 inches by cutting into a cardboard box. Note how well children visualize before selecting an object to try.

4. I can see a box-shaped object in this room. What object can I see?

Use whatever shape you want for this question. It allows you to see any misconceptions children have about the attributes of a particular shape. For example, some children may think a box shape is only a cube.

5. In a bag I can feel that an object has flat faces, sharp corners, and straight edges. What might this object be?

Do children understand the spatial terms that are used?

6. I traced around one of the faces of an object. The shape I drew was a circle. What might the object have been?

The focus of this question is to help children distinguish between a three-dimensional object and its faces.

7. Kwang used three rolling shapes, four boxes, and one cone to build something. What did she build and what might her construction look like?

Let children use suitable three-dimensional objects to construct this. Look at the variety of responses.

8. From the top, a stack of blocks looks like this diagram:

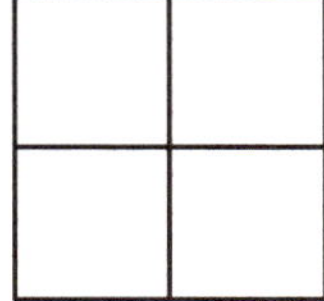

Use a set of blocks to show what the stack might look like.

Most children will make a stack with a flat top. Point out that the top does not have to be flat and then let children explore this further.

9. What do you know and what can you find out about a pyramid?

The main purpose of this question is for children to become aware of what they know. They should realize that even in common figures there are many mathematical concepts. Their responses can include comments on: the number of sides; the number of faces; the shape of the base; whether they can be stacked; and so on.

10. When I looked at a photograph taken from an airplane, I saw rectangular and circular shapes. What might these shapes be?

It is sometimes difficult for children to visualize this, although those who have been in an airplane should have a good idea. If possible, show them some aerial views to help them. You could also build a city or something similar with blocks and let children view it from above.

11. At the supermarket Mom bought a container shaped like a rectangular prism but the label came off. What might have been in the container?

This might be a question that children can do at home or when they next visit a supermarket. Keep a class list of suitable items. It can be added to as children find containers.

12. These shapes were sorted into two groups. What might the groups be?

(See Reproducible 7, Mixed Three-Dimensional Shapes)

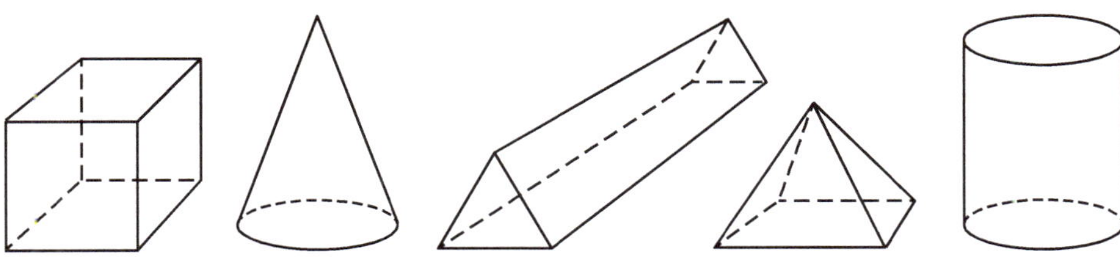

It is likely that there will be a wide range of responses. It is important to ask children to justify their answers.

13. We went on a walk. The plan was for everyone to write down the names of objects we saw and the name of the shape of that object. I wrote down the names of some of the shapes we saw but I forgot to write the name of the object next to its shape name. Can you help me work out what they might have been?

cylinder	
triangular pyramid	
cone	
rectangular prism	
sphere	

This question will identify if children know the characteristics of these shapes. Draw this table on the board. Can they identify more than one object for each shape? Give them time to share their findings.

Two-Dimensional Shapes
(Grades 3–5)

Experiences at This Level Will Help Children To

- use suitable spatial language to describe and compare shapes
- copy and make patterns that involve translations, reflections, and rotations
- explore transformations of shapes
- identify, construct, and order angles using direct comparison and appropriate instruments
- construct two-dimensional shapes with a variety of materials
- recognize and draw straight, curved, and parallel lines
- use perspective to represent three-dimensional shapes as two-dimensional
- give clear descriptions of shapes
- produce complex symmetrical patterns
- construct tessellations and explain why shapes will or will not tessellate
- enlarge and reduce two-dimensional figures
- identify and classify angles
- construct common shapes
- identify and describe horizontal, vertical, and diagonal lines

Reproducibles are available in a downloadable, printable format. See page xx for directions about how to access them.

Materials

- pattern blocks
- a clock face with moveable hands
- geoboards and rubber bands
- measuring equipment
- protractors
- small mirrors
- drawing paper
- Walking Path Shape (Reproducible 8)

Good Questions and Teacher Notes (pages 106–110)

1. On a page draw five lines, no two of which are parallel.

Ask children to check each other's drawings to see if they agree.

2. I am thinking of a shape. It tessellates, which means many of them fit together like tiles, with no spaces. What might the shape be?

If you have lots of small shapes, children could check their own answers by seeing if they fit together. Some children may need to use these shapes to answer the question. Some shapes that tessellate are triangles, squares, rectangles, hexagons, and some irregular shapes.

3. Using four triangles from the pattern blocks, make as many different shapes as you can. Draw them.

Ask children to devise a way to check if their shapes are the same or different. (If you do not have access to pattern blocks, cut out four same-size triangles for each child.)

4. A lot of things in our room have angles that are the same as a right angle.

What can you see that looks like this?

Let children use the corner of a sheet of paper.

5. I was watching television and I noticed that the hands of a clock made an acute angle. What time might it have been?

Children might like to use a clock to do this. As they are working, check that they know what an acute angle is.

6. Rohan made five squares on his geoboard, all of which were different sizes. What might Rohan's geoboard have looked like?

Children should work with a partner and share a geoboard. As they complete the five squares they should draw them on paper before beginning a new set of squares. Allow time for pairs to share their geoboard patterns with others.

7. If I took two steps forward, turned right 60 degrees, took another two steps forward and turned right 60 degrees, and kept doing this, this is the path I would walk.

(See Reproducible 8, Walking Path Shape.)

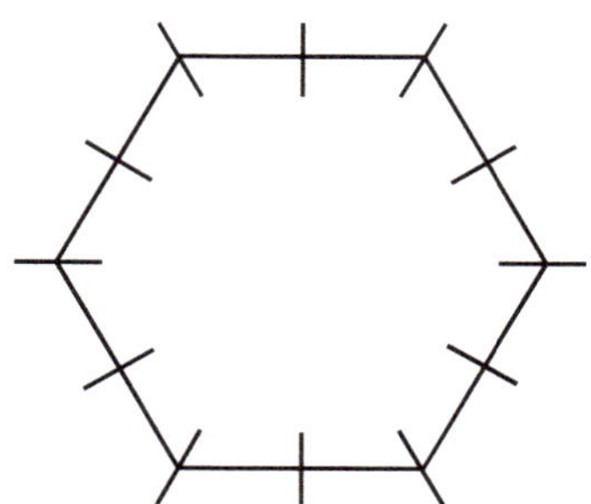

What are the instructions for some other shapes?

Let children check each other's work by following the instructions and seeing if they make the shape they say it does.

8. The fourth-grade children said they were going to mark out a basketball court. Write down the instructions they need to follow.

You would expect the children to describe the length of the lines, the angles, the size of the circles, and so on, so that a fourth-grade child could understand. Let them check each other's instructions by actually marking out a court.

9. I cut out a shape, folded it in half, and showed the new folded shape to my teacher. She said it had a line of symmetry. What might my original shape have looked like?

Allow children to make some shapes and fold them. Can they explain symmetry? You could display the finished shapes.

10. Tan told me he drew a shape that had no diagonals. What might his shape look like?

Do the children realize that their shapes must be curved or triangular?

11. If I drew a shape and the total of the angles in the shape was 180 degrees what could the shape be?

Children can draw any triangle they like. They may like to use a protractor to check the size of the angles.

12. I wrote a capital letter so that it had rotational symmetry. What might the letter have been?

Rotational symmetry means you can turn the letter and it still looks the same after a rotation of less than one full turn (e.g., S and H).

13. A shape has at least two sides 2 inches long and at least two sides 4 inches long. One angle is 90 degrees, one angle is less, but the rest are more. What might the shape look like?

After children have completed some shapes, let them check in pairs that their shapes follow the instructions.

14. A shape is made of two smaller shapes that are the same shape and the same size and that are not rectangles. What might the larger shape look like?

Check if children have followed directions. Can anyone work out how to check that the smaller shapes are congruent?

15. My friend was sure that she had made a rectangle. The teacher said it was not a rectangle. How can my friend check if it is a rectangle?

Children should be looking at the size of the angles as well as the length of the lines. Would the definition that children produce allow a square to be a rectangle?

Three-Dimensional Shapes (Grades 3–5)

Experiences at This Level Will Help Children To

- recognize three-dimensional shapes from drawings or photos taken from various perspectives
- identify and draw different cross sections of three-dimensional shapes
- make common solids from clay
- draw and make nets of common solids, and match shapes with nets
- use increasingly accurate language to describe objects
- explore repetitions of objects in structures
- represent three-dimensional shapes on a two-dimensional surface and select objects to match three-dimensional representations
- make complex models, including those that use combinations of three-dimensional objects
- classify and compare objects using all aspects of their features and properties
- investigate translations, rotations, and reflections in objects and formations
- enlarge simple three-dimensional objects

Reproducibles are available in a downloadable, printable format. See page xx for directions about how to access them.

Materials

- photographs of aerial views
- various containers and packaging from packaged foods
- fruit and vegetables
- blocks
- plastic straws and joiners or clay
- overhead projector
- squares made from light cardstock and squared paper
- cubes
- Block Diagram (Reproducible 9)

Good Questions and Teacher Notes (pages 113–114)

1. Julio and David built a building using thirty bricks. What might their building look like?

Children must use the blocks to construct one building, not a group of buildings. Make sure they have time to see the variety of possible responses.

2. From the top a stack of blocks looks like this:

(See Reproducible 9, Block Diagram.)

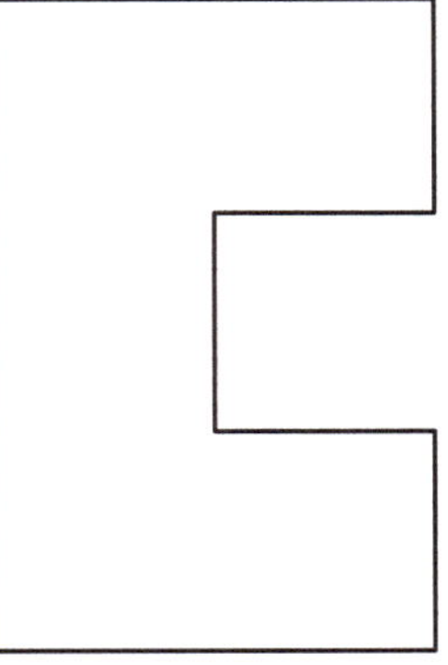

What might it look like from the front?

This activity is similar to Question 8 in the "Grades K–2: Three-Dimensional Shapes" section. Again, some children will not realize that the top does not have to be flat, and this will affect their view of the front. Likewise, the front does not have to be even.

3. I have twelve plastic straws and eight joiners. What model's skeleton can I make?

Have straws and joiners available for children to use. You may use clay to join straws. Display the finished models and discuss their diversity.

4. Six squares were joined together with one whole side of each touching at least one whole side of another square, but they could not be folded to make a box. What might the arrangement of squares look like?

As well as allowing children to join six squares together to try to make a box, also allow them to cut up boxes in different ways so that the six surfaces join along their sides (or edges). Another approach might be to give children squared paper and have them draw six adjoining squares, then cut them out to check whether they can be folded into a box. Display finished sets of adjoining squares.

5. A stack of cubes was made, glued together, and then painted on the outside. Eight of the cubes were painted on exactly two sides. What might the stack look like?

The shape does not have to be regular. The children may have to discuss what exactly two sides means.

CHAPTER 10

Good Questions for Data Analysis and Probability

GRADES K–2 116

Chance *116*

Data *120*

GRADES 3–5 123

Chance *123*

Data *127*

The two topics included in this chapter are:

1. chance
2. data

As many situations involve chance, it is important that we provide activities that will develop children's understanding of chance and allow them to handle chance situations confidently. The experiences provided by these questions range from allowing children to make predictions about the likelihood of an event occurring to developing their ability to identify and record all outcomes in a systematic manner.

The data topic provides children with opportunities to collect, organize, represent, and interpret data from situations that are both relevant and of interest to them. They are also encouraged to determine the appropriateness and quality of data collection and presentation so that they may develop a discerning approach to interpreting data.

Children's confidence and competence in working with chance and data will be developed by working on these questions.

Chance (Grades K–2)

Experiences at This Level Will Help Children To

- recognize that some events involve chance
- use and understand chance expressions
- recognize that different results are possible when the same event is repeated
- classify events as certain, possible, or impossible
- compare possible events
- distinguish impossible from unlikely
- predict results and compare with the outcome

Materials

- coins
- dice

Good Questions and Teacher Notes (pages 117–119)

1. If two coins are tossed, what could happen?

It is important that children actually toss two coins many times so that they can observe what happens. The main point is for them to see that there are three different outcomes for the one event and that any of the three can happen. Some of them may realize that the heads/tails combination occurs more often and start to think about why this is.

2. I overheard my mother telling our neighbor that on the weekend we would definitely do something but I couldn't hear what it was. What might it be?

Children's responses will indicate if they understand the meaning of *definitely*. What is definite for some will not be so for others, so it is good for them to hear what others think and recognize this.

3. I heard the teacher say, "It is . . . that all of the children in this class will watch television tonight" but I didn't hear one of the words. What might the missing word be? What is something that is more likely to happen than what the teacher is talking about?

Children may need to discuss what the question is asking them to do. Ask them to explain it in their own words. Perhaps some of them could discuss how often they watch television.

4. Two children were playing a dice game. One child tossed two dice together and when they landed one was a 6 and one was a 4. What other number combinations might the child have tossed?

Note whether children give a range of responses.

5. Someone asked the teacher a question and she replied, "Maybe." What might the question be?

This is similar to Question 2 and the way children interpret "maybe" depends on previous experiences with the word. It is worth discussing what "maybe" means to different children, or what it means when certain people say it—for example, if Mom says it, it might mean no, whereas if Dad says it, it generally means yes.

6. When we were playing a dice game where we had to throw a 6 on one of the dice to start, Emma said, "Let's make it that we have to throw a 3 instead of a 6 because it is easier." Do you agree with her? Why?

This will allow you to see if children have any misconceptions about the chances involved in tossing a dice.

7. A family has three children. We know that at least one of the children is a girl. Draw what the family might look like.

Can children give all possibilities?

8. Our class wrote down some things that we felt were impossible. What might we have written?

You can use a similar question for "certain" too. It is important to discuss what children have written.

9. Madeleine threw two dice and when they landed she subtracted one number from the other and wrote down the answer 1. What might the numbers on each have been?

Can children give all possible outcomes? Allow them to use dice if they wish.

Data (Grades K–2)

Experiences at This Level Will Help Children To

- decide what data to collect and how to collect it to answer questions
- represent data concretely and pictorially
- compare information by counting
- sort and sequence data
- make simple pictographs and block graphs using one-to-one correspondence
- describe results from data collection and display
- interpret visual representations

Reproducibles are available in a downloadable, printable format. See page xx for directions about how to access them.

Materials

- materials to make graphs (e.g., blocks, beads and string, etc.)
- Mystery Bar Graph (Reproducible 10)
- Mystery Line Plot (Reproducible 11)

Good Questions and Teacher Notes (pages 121–122)

1. What might this be the graph of?

(See Reproducible 10, Mystery Bar Graph.)

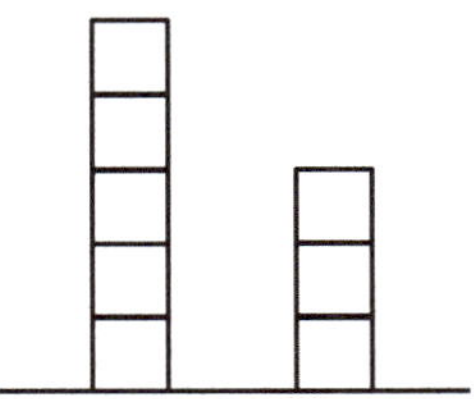

Do children make reasonable suggestions? Are some suggestions more plausible than others?

2. You did a survey to find out who was happy and who was sad. What pictures could you use to represent these feelings?

Make a display of children's suggestions.

3. The twenty-five children in our class each drew a self-portrait. Our teacher asked a question that we could answer with a "yes" or a "no." Twenty children put their picture in the Yes column while five children put their picture in the No column. What might the question have been?

Are the suggestions made by children reasonable? Make a class chart of possible questions.

4. On a graph about pets owned by children in our class, I counted more dogs than cats. What might the graph look like?

Children can represent this how they want. Do they realize that the graph may show other pets as well? Allow time for them to explain their graphs to others.

5. I did a survey of my third-grade class. This is what the graph looks like:

(See Reproducible 11, Mystery Line Plot.)

```
| X
| X
| X X
| X X X
| X X X X
| X X X X X
+-----------
  1 2 3 4 5
```

What might the survey be about?

Are children's suggestions reasonable? Discuss why some suggestions are more plausible than others. The class might like to do a survey to check whether some of the suggestions are possible or reasonable.

Chance (Grades 3–5)

Experiences at This Level Will Help Children To

- identify and record all possible outcomes from simple chance experiments
- identify some outcomes as being equally likely
- use simple techniques for random selection
- order events from most likely to least likely and justify choice
- realize how an outcome can be influenced
- appreciate the idea of fair and unfair in simple games
- analyze outcomes from simple chance experiments
- use appropriate language of chance
- use a numerical scale for chance events
- interpret probability statements
- design a simple random device to produce a specified order of probability

Reproducibles are available in a downloadable, printable format. See page xx for directions about how to access them.

Materials

- dice and/or cubes to make dice
- a bag and tiles, at least two colors
- materials to make board games
- playing cards
- Blank Spinner Faces (Reproducible 12)
- blank cards to write numbers on

Good Questions and Teacher Notes (pages 124–126)

1. In a bag there are some tiles. I draw out one tile and it is red. I put it back and draw again. This time the tile is blue. I put it back. After ten draws, I have drawn out three red and seven blue. How many tiles might there be in the bag and how many might be blue?

There might be any number of tiles in the bag, but we suspect that they are mainly red and blue. The proportions do not have to be 3 to 7, but it is likely that there are more blue than red. Children might like to do some experiments.

2. My older sister was talking to Dad and asked him a question. His reply was, "It is more likely than unlikely." What might the question be?

Children should be asked to justify their question. This activity focuses on the language of chance.

3. The probability of an event is $\frac{1}{3}$. What might the event be?

Can children offer a range of suitable responses? You may have to discuss what $\frac{1}{3}$ means. Children should justify their responses.

4. What words could be used to describe an event with a probability of 0.6?

Again, this checks if children can interpret numerical statements of probability. Some possible words are "good chance," and "better than even chance."

5. A toy manufacturer wants to design a die where the chance of throwing a "red" is $\frac{1}{2}$. What might the dice look like?

As long as half of the faces are red, it does not matter what the others are. Children could make their dice and test them out.

6. Using a small deck of cards you have a $\frac{1}{4}$ chance of choosing a picture card and $\frac{1}{2}$ chance of drawing an even number. What might the cards in the deck be?

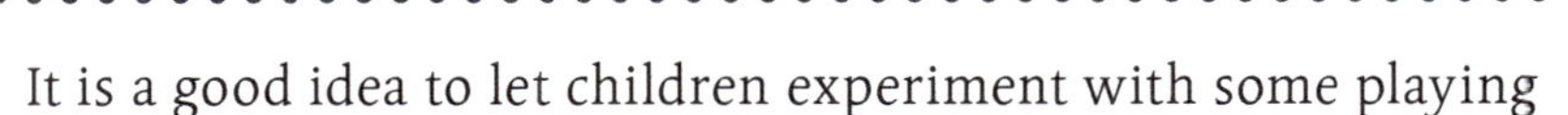

It is a good idea to let children experiment with some playing cards to help them do this one.

7. We spun a spinner lots of times. It landed on blue most of the time, on red some of the time, and only once on white and yellow. Draw what the spinner face might look like.

(See Reproducible 12, Blank Spinner Faces.)

Let children color the spinner faces in how they think and then try the spinners using a paper clip and a pencil mounted in the middle to see what results.

8. There were four cards in a container. One card had the number 2 on it, another had 5, another had 4, and the last had 7. I had to select two of the cards at a time. On my first try I selected 5 and 7. What other combinations might I have selected?

Do children record the combinations systematically? Do they record all possible combinations?

Data (Grades 3–5)

Experiences at This Level Will Help Children To

- collect data to answer questions
- plan appropriate and efficient ways to organize data
- improve descriptions of categories
- use many-to-one correspondence to display data
- represent data on bar graphs and use scales on axes
- interpret information from tables and graphs
- suggest what data to collect to answer questions
- prepare questionnaires where necessary to collect data
- record data systematically
- represent and interpret data on a wide range of graphs, tables, and diagrams
- use a database to enter and extract information
- use fractions to summarize data
- describe trends in line graphs
- make judgments about the suitability of data representations

Reproducibles are available in a download-able, print-able format. See page xx for directions about how to access them.

Materials

- materials to make graphs (e.g., blocks, beads and string, etc.)
- calculators
- Favorite Shows Pie Chart (Reproducible 13)
- Mystery Pictograph (Reproducible 14)
- Hunger Graph (Reproducible 15)
- Mystery Survey Results (Reproducible 16)
- Mystery Pie Chart (Reproducible 17)
- Sports Pie Chart (Reproducible 18)
- Animal Pictograph (Reproducible 19)
- Measurement Line Plot (Reproducible 20)
- Graph of Children Talking (Reproducible 21)

Good Questions and Teacher Notes (pages 128–134)

1. This graph shows the proportion of children in a class who prefer particular shows.

(See Reproducible 13, Favorite Shows Pie Chart.)

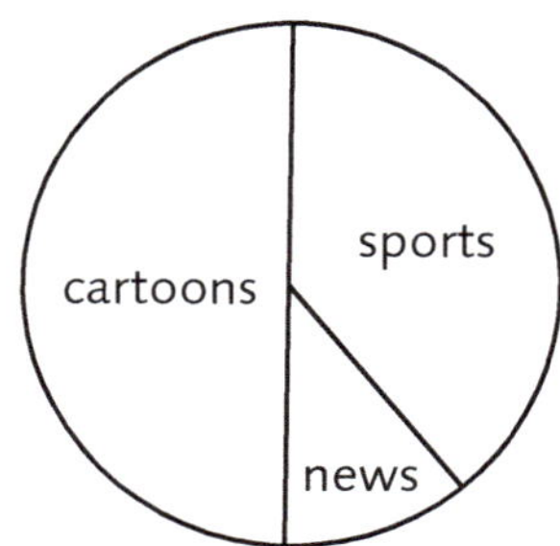

How many children might be in the class and how many prefer each show? Present the data a different way.

Note how children approach this. Do they decide how many children are in the class first or do they try to work out how many children would be in each category first? Do they realize that there must be an even number of children in the class? Allow time to share their presentations and discuss the most effective ways to present the data.

2. This is a graph about a fourth-grade class. What is it about? How many children might there be in the class?

(See Reproducible 14, Mystery Pictograph.)

The main point here is that each symbol might represent more than one child. Do children realize this and do they give reasonable responses?

3. This is a graph of how hungry you are. What times of day might be represented by the graph?

(See Reproducible 15, Hunger Graph.)

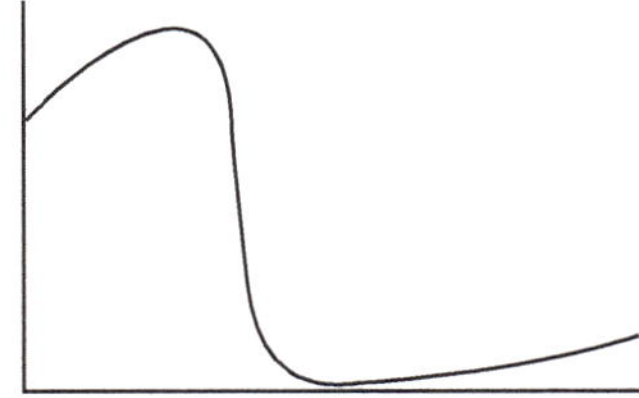

Can children relate points on the graph to appropriate times of day? They might like to discuss what the graph would look like near meal times, and before and after meal times.

4. Below are the results of a survey. What might the survey be about?

(See Reproducible 16, Mystery Survey Results.)

Place	Tally
library	///
playground	~~////~~ ~~////~~ //
under the trees	~~////~~ //
driveway	///

Accept any reasonable suggestions. For example, it could be a survey of where children like playing, a survey to show where accidents have happened, and so on.

5. A graph showed that among a class of children the most popular footwear was sneakers. The next most popular was boots. The next was sandals and the least popular was flip-flops. What might the graph look like?

This question allows children to interpret the data in their own way. Look for those who display data using many-to-one correspondence. Display finished graphs.

6. In a survey of this class exactly half the children said "yes" and half said "no." What might the survey be about?

Are children's responses reasonable? You might like to do a survey with the class to check whether the suggestions are reasonable.

7. What might this be the graph of?

(See Reproducible 17, Mystery Pie Chart.)

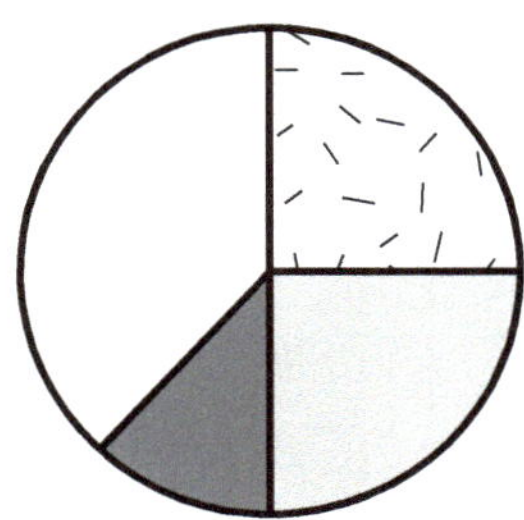

Are children's suggestions reasonable? Discuss why some suggestions are more plausible than others.

8. This is the result of a survey of children in fifth grade at a school similar to this one. Show the data in some different ways.

(See Reproducible 18, Sports Pie Chart.)

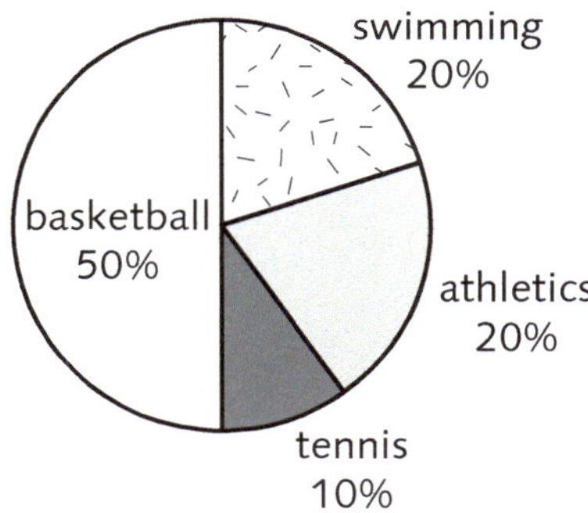

Note how children approach this. Do they decide how many children are in the class first? Do they realize that there must be an even number of children in the class? Allow time to share their presentations of alternative ways to present the data and discuss which are the most effective.

9. We did a survey of a class. Below are the results. How many children might be in the class? Represent the data in a different way.

(See Reproducible 19, Animal Pictograph.)

Ask children to justify their figure for the number of children in the class. Do they realize that their representations will depend on whatever scale they choose to use? Are they aware that each symbol might represent more than one animal?

10. For homework, students measured a common household object in their homes. Their data is recorded here. What might have been the object? What do you notice and what do you wonder about the data set?

(See Reproducible 20, Measurement Line Plot.)

8	$8\frac{1}{4}$	$8\frac{1}{2}$	$8\frac{3}{4}$	9	$9\frac{1}{4}$	$9\frac{1}{2}$	$9\frac{3}{4}$	10	$10\frac{1}{4}$	$10\frac{1}{2}$	$10\frac{3}{4}$
							X				
					X		X				
				X	X		X				
				X	X		X				
	X	X		X	X	X	X				
	X	X		X	X	X	X			X	

Be sure to discuss possible reasons for the differences in measurements including that there are likely different sizes of the same object (e.g., scissors) as well as differences in accuracy.

11. This is the number of children talking in the class over a period of thirty minutes. What time of day might it be?

(See Reproducible 21, Graph of Children Talking.)

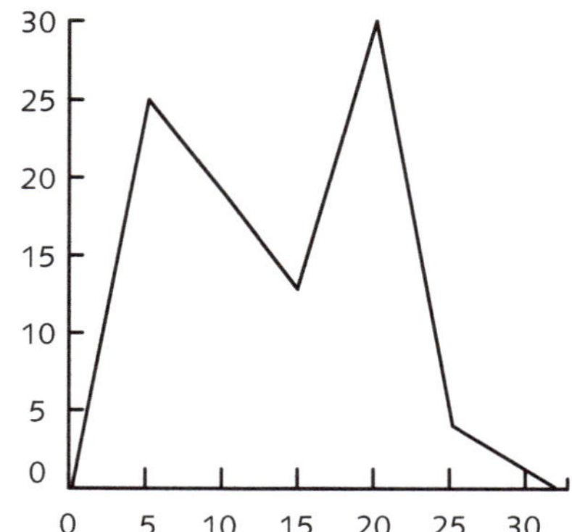

Before starting, discuss with the class how they interpret the graph. After working on the task, ask children to justify their responses. Can they give more than one response?

CHAPTER 11

Good Questions for Measurement

GRADES K–2 136

Weight 136
Time 140
Length and Perimeter 145

GRADES 3–5 149

Weight 149
Volume 152
Area 158
Time 165
Length and Perimeter 168

The five topics included in this chapter are:

1. weight
2. volume
3. area
4. time
5. length and perimeter

We use measurement in most aspects of our daily lives, whether it be to estimate, interpret, or measure quantities accurately.

Children's experiences in answering these questions will enable them to develop an understanding of the importance of measurement while becoming familiar with the various concepts involved.

They will understand the need for more precise measurements and be able to choose appropriate units and instruments for their measuring. They will recognize situations where a reasonable estimate is more appropriate than an exact measurement.

You will find many links between the various topics. However, the questions within each topic have their main teaching point within that topic. You can adapt questions by changing the amounts or quantities to suit the children in your class.

Weight (Grades K–2)

Experiences at This Level Will Help Children To

- use suitable language of comparison for weight (e.g., heavier, lighter, too heavy, too light)
- make estimates based on their own capability to lift, pull, push, and so on
- estimate order and compare weights by feel and by a pan balance
- choose appropriate attributes for ordering
- work to improve judgments of weight
- use direct and indirect comparison to compare and measure weight
- recognize the need for common units when direct comparisons are impossible
- choose appropriate uniform units to measure and compare

Materials

- various objects for weight activities, including: a pen, a small bucket of water, a book, a potato, a lunchbox, bottle tops, jars, rice, and various other objects
- pan balances
- weights (e.g., objects that weigh 1 pound)
- coins

Good Questions and Teacher Notes (pages 137–139)

1. What can you find that is lighter than a pen?

2. Find something that you can lift that is heavier than this bucket of water.

Questions 1 and 2 focus on language and comparing by lifting. It is important that children get the opportunity to talk about what they did.

3. A book is on one side of a pan balance, and two objects are on the other side so the pans are level. What might the two objects be?

Children need a pan balance to do this. Do they realize that different books have different weights?

4. What can you find that is bigger than a potato but lighter than it?

5. What can you find that is heavy but small?

6. Lin carried a big, full bucket quite easily. What might have been in the bucket?

Questions 4, 5, and 6 require children to focus on the two attributes of weight and size. Again, the language used, both by the children and the teacher, is essential to the development of concepts of measurement.

7. Using a pan balance I balanced my lunch box with some bottle tops, but I cannot remember how many. How many might I have used?

Answers will vary depending on what type of lunch box you use and what is in it. They also depend on the bottle tops used. Note children who are aware of these variations.

8. André carried an object in a bag. When he put another object the same as the first object into the bag he could not lift it without struggling. What might the objects be?

Note if children make reasonable suggestions.

9. What objects can you find in your home that are approximately 1 pound? Ask someone at home to help you make a list.

Allow children to report on what they found.

10. Fill a small jar with rice. Now find another container and fill it with water so it is the same weight as the jar with rice.

Children should see that objects that have a different size, shape, and texture can have the same weight.

11. Can you find two objects that have the same size but different weights?

This question focuses on two attributes: size and weight. Are there any children who cannot distinguish them?

12. I can see an object that weighs more than 1 pound but less than 2 pounds. What might the object be?

Do children choose suitable objects to measure? Do they have a good idea of what 1 pound might feel and look like?

TIME (GRADES K–2)

EXPERIENCES AT THIS LEVEL WILL HELP CHILDREN TO

- use and respond to language that compares and describes time
- order daily activities and sequence on a simple timeline
- learn names of days and months, and sequence months and seasons
- relate days and months to events in their own life
- understand that clocks are used to tell the time and that analog and digital clocks provide the same information
- order times of day or year by natural or cultural events
- know the function of a calendar and locate dates and events on it
- estimate time of day
- recognize o'clock and half past times on an analog clock
- tell time on an analog and digital clock in hours and minutes
- estimate time of day, week, or year using obvious indicators
- measure with standard units using a variety of timers
- classify events according to duration of time
- calculate times before or after given times (minutes and hours)
- make and read simple schedules
- understand a.m. and p.m.
- know simple time facts (e.g., sixty seconds is equal to one minute, seven days is equal to one week)

MATERIALS

- calendars
- analog clocks with movable hands, including the second hand
- a picture or photo of an event that happens during a school day
- digital clocks

GOOD QUESTIONS AND TEACHER NOTES (PAGES 141–144)

1. What is something we could do that takes exactly one minute?

Rather than tell children what to use or do to measure a minute, it is interesting to see what they think is useful, and how they go about this. Their responses will tell you if they understand how long a minute actually is. If they say something like, "count to sixty" or "walk down the hallway and back again," then you can be confident they have a good understanding. If they say something like, "walk to the supermarket and back again" or "eat breakfast," then you can be reasonably sure their understanding is lacking.

2. What are some things you do in the morning and some things you do in the afternoon?

Children's responses will indicate if they know the distinction between these two time periods.

3. I know two people who have their birthday in the month before yours and two who have it in the month after yours. Who might they be?

Let children hear everyone's responses. This will help them to learn the sequence of months.

4. Show children a picture of something that happens during the day. Ask, "What are some things that happen before/after this picture?"

This focuses on the ordering of daily events into a sequence.

5. What things could you do that take about one hour?

This is similar to Question 1 but does not require the exact measurement of one hour.

6. My mom said that I went to bed later than my usual bedtime of eight o'clock. What time might I have gone to bed?

Note if children understand the concept of "later" as far as clock time goes.

7. What are some things you do each school day between twelve o'clock and four o'clock?

Are children able to relate daily events to clock time?

8. Show a time on an analog clock face with movable hands. Tell what you might be doing at that time on a school day.

See what children know about clock time.

9. There is something you do after you get out of bed and before you go to school that takes approximately four minutes. What might it be?

Note if children can give reasonable suggestions.

10. I left home and arrived at school forty-five minutes later. When might I have left home and when might I have arrived at school?

Note if children are able to calculate times before or after a certain time.

11. The hands of a clock make an angle that is less than a quarter of a turn. What time might it be?

Children who can work out an answer without using a clock face have strong visual ability. Let children use a clock face with movable hands if they want. Do children systematically record answers?

12. I am a month with thirty-one days. Which month might I be?

Do children list all possibilities?

13. Raphael took exactly thirty seconds to do each of three things. What might the three things have been?

Children should be able to give reasonable suggestions to indicate that they know how long thirty seconds is.

Length and Perimeter

(Grades K–2)

Experiences at This Level Will Help Children To

- use and respond to appropriate language of comparison for length
- compare and order lengths by using direct and indirect measures
- measure using nonstandard units
- recognize the need for a common unit when direct comparison is not possible
- compare and estimate distances using everyday language
- be aware that perimeter is the length of a shape's boundary
- distinguish different length attributes (e.g., length, width, height)
- realize the need for standard units
- select a suitable unit to measure and compare lengths and perimeters
- realize the importance of accuracy
- measure and record length
- improve estimates of length and perimeter

Materials

- interlocking cubes
- lengths of card or paper, sticks, string
- rulers, both customary and metric
- Cuisenaire rods

Good Questions and Teacher Notes (pages 146–148)

1. Find as many objects as you can that are longer than 3 handspans but shorter than 4 handspans.

This question encourages children to use comparison language, for example, almost as long as, exactly the same as, and so on.

2. What is longer than 2 of your foot lengths but shorter than 3 of your foot lengths?

Observe how children do Questions 1 and 2. Do they line their handspans or foot lengths up against the object to be measured (direct)? Do they mark the length off on a card or piece of paper and then use this as their measurer (indirect)? Do they use a formal measuring instrument, like a ruler, to find how long 3 handspans or 2 foot lengths are and then use the ruler to measure other things? This last method is the most sophisticated. Children who use this should not be forced to use other methods unless they choose to.

3. Can you find something that is the same length as your height?

Do children only measure vertical objects or do they understand that height is a length measurement? Discuss why things vary between children; for example, some are taller than others.

4. Give children a stick or piece of string. Ask, "Can you find some things longer than/shorter than/the same length as your string?"

This question establishes if children understand the concepts of longer/shorter/same length as.

5. There are five children standing together. If you are the middle person in height, who might the other four be?

This activity requires children to compare and order heights. Note how they do this; for example, directly or by using an instrument.

6. Sharbat measured a table and said that it was 10 sticks long. Michael measured the same table and said it was 12 sticks long. How might this happen?

Some possible reasons are: the two sets of sticks they used were different; Sharbat did not put her sticks exactly end to end; Michael did not measure in a straight line; Sharbat did not start at the edge of the table; the sticks used by either child varied in length.

7. What in this room is longer than 1 foot but less than 2 feet?

Children should be able to estimate things between 1 foot and 2 feet in length.

8. Can you find something that is about twice as long as it is high? What is the length of each?

Note if everyone understands the language used (e.g., *long*, *high*). Check how confident children are with doubling and halving.

9. Give children a piece of string. Say, "This string is the distance around some objects. What might some of those objects be?"

Watch what children choose to compare the string against. Are they choosing objects that have a perimeter about the same length as their string? Do their choices improve with practice?

10. Find as many things as you can that are one inch long. Then find as many things as you can that are one foot long.

Note if children choose suitable objects to measure and if they use a ruler correctly. You can modify this question by changing the system of measure from customary to metric.

Weight (Grades 3–5)

Experiences at This Level Will Help Children To

- choose and use appropriate standard units and instruments when measuring and comparing objects
- recognize that a smaller unit will give a more accurate measurement
- read scales accurately
- estimate by using known weights of objects and standard measures

Materials

- scales (e.g., spring scale)
- various objects, some with a weight less than 1 pound and some between 1 pound and 2 pounds
- marbles, fruit

Good Questions and Teacher Notes (pages 150–151)

1. Seung weighed a bag of flour and found it to be $5\frac{1}{4}$ pounds. Pedro weighed the same bag and found it to be $5\frac{1}{2}$ pounds. How could this happen if they used a balance scale? How could this happen if they used a spring scale?

The balance scale may not have been level; some flour might have leaked; the flour might have been unevenly placed. For the spring scale there can be zero errors, reading errors, or inaccurate scales.

2. Can you find a collection of objects with a total weight of 5 pounds?

Check how easily children can convert ounces to pounds. Do they choose objects whose weights are near to the desired weight?

3. A jar of marbles weighs 1 pound. How much might each marble weigh? How much might the jar weigh?

Note how children work this out. Do they assume all the marbles are the same size or do they imagine different size marbles in their jar?

4. I bought 1 pound of fruit. What might I have bought and how much might each piece weigh?

Children will need various pieces of fruit to find their weights. Do they use a calculator to help them calculate how much of each type of fruit makes 1 pound? If not, how do they work this out?

5. A school bus has a 5,800-pound carrying capacity. How many people would it be permitted to carry?

When working this out do children distinguish between adults and children? Discuss how bus companies work out how many people may travel on a bus. Do any children make allowances for baggage?

6. Make a list of some objects in your house that weigh between $\frac{1}{2}$ pound and 1 pound.

Allow children to report on what they find. Is one weight more common than others?

Volume (Grades 3–5)

Experiences at This Level Will Help Children To

- use suitable language of comparison for volume and capacity
- fill containers by packing and pouring
- choose and use nonstandard units of measure when comparing the size of two containers
- use nonstandard units to measure capacity
- compare two containers by pouring from one to the other
- estimate volume and capacity and improve estimates by comparing with known sizes of common objects
- use uniform materials (e.g., cubes, to measure and compare volume and capacity)
- choose appropriate units and instruments to measure, compare, and describe capacity and volume
- compare and order capacity in common standard units
- make estimates and describe them appropriately
- use smaller units for accuracy
- read various measurements of capacity accurately

Reproducibles are available in a downloadable, printable format. See page xx for directions about how to access them.

Materials

- assorted jars, cups, and other containers
- matchboxes, other boxes of various sizes and shapes
- assorted soft drink and juice containers
- water, rice, sand, and so on
- cubes (interlocking and noninterlocking)
- light cardboard and paper to make models
- Cubic Structure (Reproducible 22)
- Rectangular Box (Reproducible 23)
- Cube Diagram (Reproducible 24)

Good Questions and Teacher Notes (pages 153–157)

1. Find as many containers as you can that will hold more than this jar.

Note how children choose the containers. Are their estimates good? Do they pour from one container to another to check?

2. Sasha filled a container using 3 cups of water. What container might she have filled?

This depends on the cup that was used. It is important to let children pour water to fill various containers so they can find possible ones.

3. Design some box-shaped buildings using exactly twenty-four cubes.

Recording is important. Ask children to share their recording strategy. Could someone else build their building from the description?

4. Can you find some containers that have the same capacity but a different shape?

Children can do this either by packing or pouring.

5. A cubic structure is made out of twenty-seven smaller cubes. Two of the smaller cubes are removed from the larger structure. What might the structure now look like?

(See Reproducible 22, Cubic Structure.)

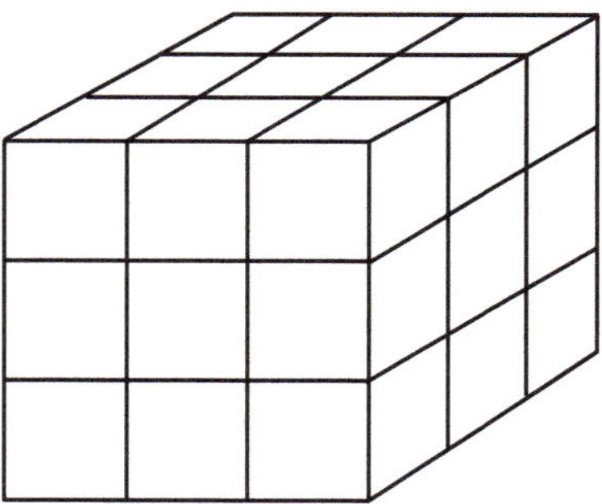

It is important to allow children to use cubes to construct this rather than expecting them to visualize it.

6. A rectangular box is made out of cubes. The end of the box looks like the below. What might the volume of the box be?

(See Reproducible 23, Rectangular Box.)

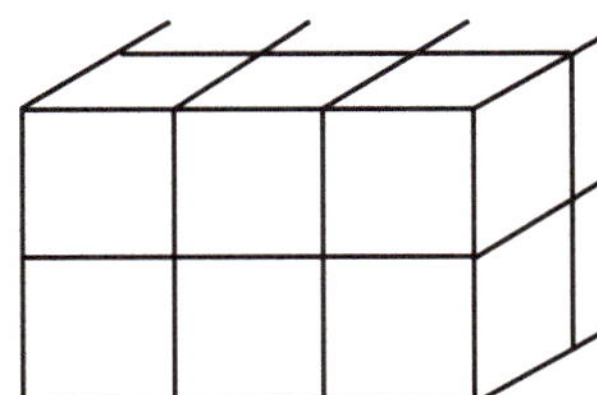

The answers are multiples of six. Let children use cubes to do this.

7. I can see a container with a capacity bigger than 1 quart but smaller than 3 quarts. What might the container be?

Some children may be able to do this by reading labels of containers (if attached).

8. Design some boxes that can hold thirty-six chocolates, each of which is a small cube.

This could be a flat tray box with the cubes arranged as 6 × 6, 4 × 9, 3 × 12, 2 × 18, or 1 × 36. It could also have more than one layer. Are children aware of the possibilities?

9. Design a building constructed out of twenty cubes. It has some sections that are more than one story high. Draw your designs.

The key feature of children's designs is whether someone else can read the design and build it.

10. I made a shape from cubes. It looks like the diagram below. What might its volume be?

(See Reproducible 24, Cube Diagram.)

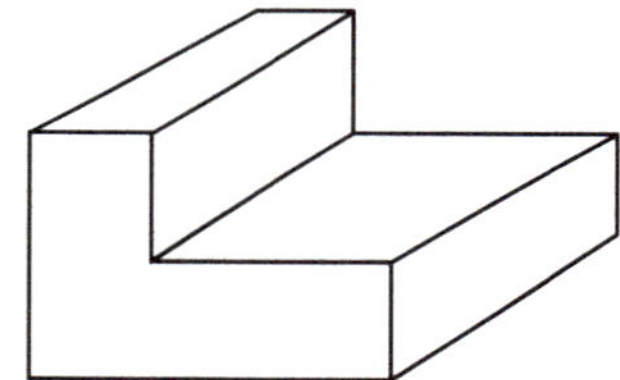

Can children see a pattern? Can they find more than one possible answer? What answers are not possible?

11. A container holds 100 teaspoons of water. What might the container look like?

Children should share their strategies, especially for finding the volume of a teaspoon, with the class. Discuss the methods they use for recording their solutions.

12. A man says he will pay $1.00 for every grain of sand he can hold in his hand. How much might this be worth?

Children should share strategies. Some important variables are the size of the grains and the size of the hand.

13. I used some cubes to make a larger cube. How many cubes might I have used?

Do children understand what a cube is? Can they use a pattern when working this out; that is, 8 cubes will make a 2 × 2 × 2 cm cube, 27 cubes will make a 3 × 3 × 3 cm cube, 64 cubes will make a 4 × 4 × 4 cm cube, and so on?

Area (Grades 3–5)

Experiences at This Level Will Help Children To

- describe surfaces by touching and looking
- cover surfaces with a variety of flat objects
- estimate order of areas and check estimates by comparing directly or indirectly
- use appropriate language (e.g., "This covers more than that.")
- select suitable uniform units when comparing and measuring
- use accuracy (e.g., leaving gaps affects the result)
- use direct and indirect comparison to compare and order areas
- improve estimates of area
- understand the importance of standard units
- find or make different things with one measurement the same and another different, for example, same area/different perimeter
- use known sizes of common things to help make estimates of area
- describe the relationship among length, width, and area
- measure and compare areas of rectangles in square units

Reproducibles are available in a downloadable, printable format. See page xx for directions about how to access them.

Materials

- 1-inch square tiles
- books
- various uniform units (e.g., bottle tops or cards)
- meter squares (make them out of newspaper joined with tape)
- centimeter-squared paper; inch-squared paper
- Letter O (Reproducible 25)

Good Questions and Teacher Notes (pages 159–164)

1. What are three things this page would completely cover?

Children need to compare the page directly against other things. This question is to develop the idea of area as covering.

2. What are some things your hand will cover so that they cannot be seen? What are some things you cannot cover completely with your hand?

This again develops the idea of area as covering and encourages children to compare areas by direct matching.

3. What are some things you could cover exactly with four books?

This focuses more on covering a surface with repeat units, that is, books. Note if children choose books of the same size or different sizes, or if they use the same four books for each surface. Do they place the books close together so there are no gaps?

4. I am thinking of a shape with an area of 30 square tiles. What might the shape look like?

If children cannot visualize this let them make some arrangements with thirty tiles to help them. They do not have to be rectangular.

5. I used twenty objects (all the same) to cover my table with no gaps. What might the objects be?

Encourage children to guess before they do this. When they are trying objects note if they choose things that fit together without leaving gaps.

6. Catori covered approximately sixty-five squares on a centimeter grid. What object might she have used to cover the squares?

This focuses on using uniform counting units to compare areas. Note how children handle parts of squares.

7. My grandmother bought a square rug and each side measured 2 yards. When she got it home it would not fit in the hallway so she cut the rug up and joined the pieces together again to make a shape that would fit. What might her rug look like now?

The focus of this question is to show that different shapes can have the same area.

8. Can you find some things that have a greater area than the top of your desk but not much greater? Can you find some things that have a smaller area than your desktop but not much smaller?

This question is to make children use methods other than locking to compare areas. Note how they do this. Do they use uniform units to find surfaces just larger and just smaller?

9. A teddy bear left a footprint on grid paper. It measured 8 squares. What might the footprint look like?

Children need to discuss how they can count the squares when the shape does not completely cover a square.

10. A footprint was drawn on grid paper. Maggie said its area was 20 square units. Ivan said that it was 19 square units. How could this happen?

The purpose of this question is to focus on the errors that can be made when calculating area by counting squares.

11. I want to make a vegetable garden in the shape of a rectangle. I have 200 feet of fence for my garden. What might the area of the garden be?

This is to show that perimeter and area are different, and that shapes with the same perimeter may have different areas.

12. Draw some rectangles that have an area of 24 cm².

Note which children find all the possibilities by recording their answers methodically.

13. Letter O has an area of 6 cm². What letters can be drawn with an area of 6 cm²?

(See Reproducible 25, Letter O.)

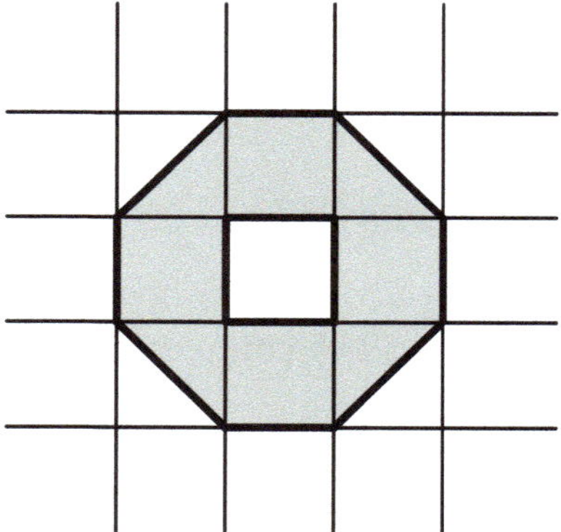

This is to emphasize that shapes of the same area can look different.

14. A rectangle has an area of 36 cm². What might its perimeter be?

Possible whole-number side lengths and perimeters (in parentheses) are 6 × 6 (24); 4 × 9 (26); 3 × 12 (30); 2 × 18 (40); 1 × 36 (74).

15. The area of a 4-by-4-foot-square tabletop is 16 square feet. By putting four tables together, what shaped tables can I make and what is the perimeter of those shapes?

Have square tiles for children to make different shapes of tables. The focus of this question is to show same area/different perimeters.

16. Sixty percent of a school property is building. One-half of the rest is asphalt. What might the area of the property be, and what might the area of the asphalt be?

This question combines fraction work with area. Note how realistic children's answers are.

17. Draw as many triangles as you can with an area of 6 square inches.

You will find some children who are able to handle this question. It shows that there are many possibilities and that height and base are important, not the sides. Let children use squared paper. Using whole number multiples, the area can be achieved with base 1 and height 6 (1, 6), as well as (2, 3), (3, 2), and (6, 1).

18. An L-shaped room has a perimeter of 20 meters. What might the area of the floor be?

Provide grid paper for children to draw their L shapes. Note any children who have difficulty drawing L shapes with perimeters of 20 units. This activity provides a very rich investigation.

19. The difference in areas of two rectangles is 32 cm^2. What might the widths and lengths of the two rectangles be?

Note how children start this one. The easiest way is to choose an area, say 100 square units, and then subtract 32 to find the other area. It is then just a matter of working out the widths and lengths of each rectangle.

20. Redesign this classroom using the same furniture as we have already. Present your design on a map or plan drawn to scale.

Children may like to use grid paper. Have them explain their designs to the class. It is important that the designs are functional.

Time (Grades 3–5)

Experiences at This Level Will Help Children To

- estimate time of day, week, or year using obvious indicators
- tell time on an analog clock
- measure with standard units using a variety of timers
- classify events according to duration of time
- calculate times before or after given times (minutes and hours)
- make and read simple schedules
- understand a.m. and p.m.
- know simple time facts (e.g., sixty seconds is equal to one minute, seven days is equal to one week)
- estimate and measure time and duration of time
- use twelve-hour and twenty-four-hour timetables
- prepare feasible timetables and appreciate the importance of time in work situations
- move easily between digital and analog representations of time
- compare and use different calendars
- investigate various time zones

Materials

- analog clocks with movable hands, including the second hand
- digital clocks
- television programming guides
- calendars
- maps of the local area

Good Questions and Teacher Notes (pages 166–167)

1. What is something you can do about one hundred times in one minute?

It is interesting to see how children work this out. Do they use the fact that sixty seconds = one minute and work in parts of a minute? Do they use trial and error? You can extend this question by changing the time (e.g., "What is something you could do one thousand times in a day?")

2. Listen to or read the following dialogue. What might the time be now?

Child: Can I go out with my friend in one hour?

Mom: It is possible it will be dark then.

Children's responses will depend on the month and Daylight Saving Time. Discuss whether they know when it will get dark today.

3. Some workers started work exactly on the half hour and worked for six-and-a-quarter hours before stopping. When might they have started and when might they have stopped?

This allows you to see if children are comfortable with calculating hours and minutes counting backward and forward. If the workers started at 6:30 they would finish at 12:45; at 7:30 they would finish at 1:45; and so on.

4. What are the times when the hour hand and the minute hand are at right angles?

Do children systematically record all the times this happens?

5. Our car trip took two and a half hours. We traveled at an average speed of 60 miles per hour. Describe our journey.

Let children use a map for this question. They might like to produce their own map to show the journey.

6. Aria was allowed to watch television for a total of five hours from Monday through Friday. What might her TV schedule be for this week?

Can children read and interpret the TV schedule?

Length and Perimeter

(Grades 3–5)

Experiences at This Level Will Help Children To

- realize the need for standard units
- select a suitable unit to measure and compare lengths and perimeters
- realize the importance of accuracy
- measure and record length
- improve estimates of length and perimeter
- recognize that a smaller unit will give a more accurate measurement
- select appropriate measuring instruments
- read measurement marks accurately
- devise and use own methods to find the perimeter of polygons
- use known lengths to help and improve estimates
- find or make different things with one measurement the same and another different (e.g., for example, same area/different perimeter)

Materials

- rulers, both customary and metric
- string
- Cuisenaire rods
- centimeter-squared paper
- inch-squared paper
- Package with Ribbon (Reproducible 26)

Good Questions and Teacher Notes (pages 169–171)

1. What could we use a yard stick or meter stick to measure?

Children should choose suitable objects to measure. They should measure horizontal and vertical lengths.

2. What can you find that has a perimeter of about 30 centimeters?

Do children understand the concept of perimeter? Are they looking at the measurement around shapes? Do they use a suitable strategy to measure perimeter?

3. Find as many different ways as you can to make a rectangle with a piece of string 3 feet long. What are their areas?

Children will find it easier to work with others to do this so they can hold the corners. Some of them may not need to use the string. Note if they record their answers methodically.

4. I have drawn a shape on centimeter-squared paper with a perimeter of 16 centimeters. What might my shape look like?

Give children centimeter-squared paper to use. Tell them that shapes do not have to be rectangles, but they should draw on the grid lines. Have them compare the areas of the shapes they draw.

5. Draw different shapes with the same area. Compare their perimeters.

Children can use either inches or centimeters. Encourage children to look for generalizations (e.g., long and skinny shapes have greater perimeters than __________).

6. I want to go on a long bike trip. I want to ride at least 500 miles but not more than 750 miles. Where might I travel?

Children need to use a detailed map with a scale rather than one with distances marked on it. They should justify their answers. They might like to discuss how they measured the distance along curved roads.

7. It takes 1 meter of ribbon to tie up a package. Assume that the bow needs 30 centimeters. What might the dimensions of the box be?

(See Reproducible 26, Package with Ribbon.)

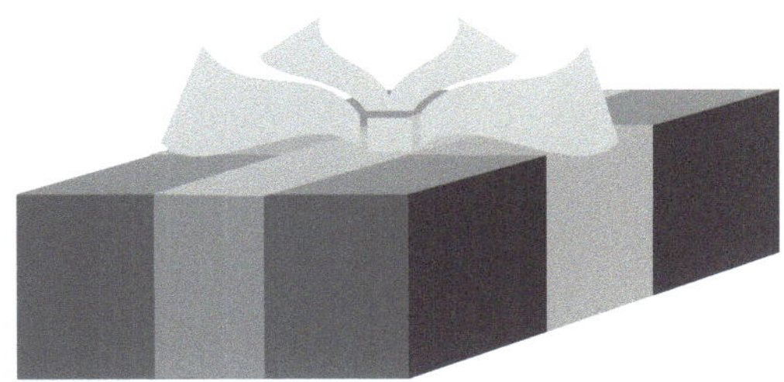

Children need to give dimensions for height, length, and width. One possible answer is 5 centimeters high, 10 centimeters wide, and 15 centimeters long.

Reproducibles

The following reproducibles are referenced throughout the text. These reproducibles are also available in a downloadable, printable format. For access, visit http://hein.pub/MathOLR and register your product using the key code **GQK5**. See page xx for more detailed instructions.

Reproducible 1	Cafeteria Price List
Reproducible 2	Money Number Cloud
Reproducible 3	Jigsaw Piece
Reproducible 4	Decimal Number Cloud
Reproducible 5	Triangles
Reproducible 6	Mixed Two-Dimensional Shapes
Reproducible 7	Mixed Three-Dimensional Shapes
Reproducible 8	Walking Path Shape
Reproducible 9	Block Diagram
Reproducible 10	Mystery Bar Graph
Reproducible 11	Mystery Line Plot
Reproducible 12	Blank Spinner Faces
Reproducible 13	Favorite Shows Pie Chart
Reproducible 14	Mystery Pictograph

(Continued)

Reproducible 15	Hunger Graph
Reproducible 16	Mystery Survey Results
Reproducible 17	Mystery Pie Chart
Reproducible 18	Sports Pie Chart
Reproducible 19	Animal Pictograph
Reproducible 20	Measurement Line Plot
Reproducible 21	Graph of Children Talking
Reproducible 22	Cubic Structure
Reproducible 23	Rectangular Box
Reproducible 24	Cube Diagram
Reproducible 25	Letter O
Reproducible 26	Package with Ribbon

The following reproducible is referenced and used throughout the book:

Reproducible A	Hundreds Chart

Reproducible 1

Cafeteria Price List

Peanut butter sandwich	$1.10	Salad	$1.55
Ham rollup	$1.40	Bag of chips	$0.65
Fruit salad	$1.15	Piece of fruit	$0.20
Cookie	$0.15		

For use with Chapter 4, Question 7, section Grades K–2: Money.

Reproducible 2

Money Number Cloud

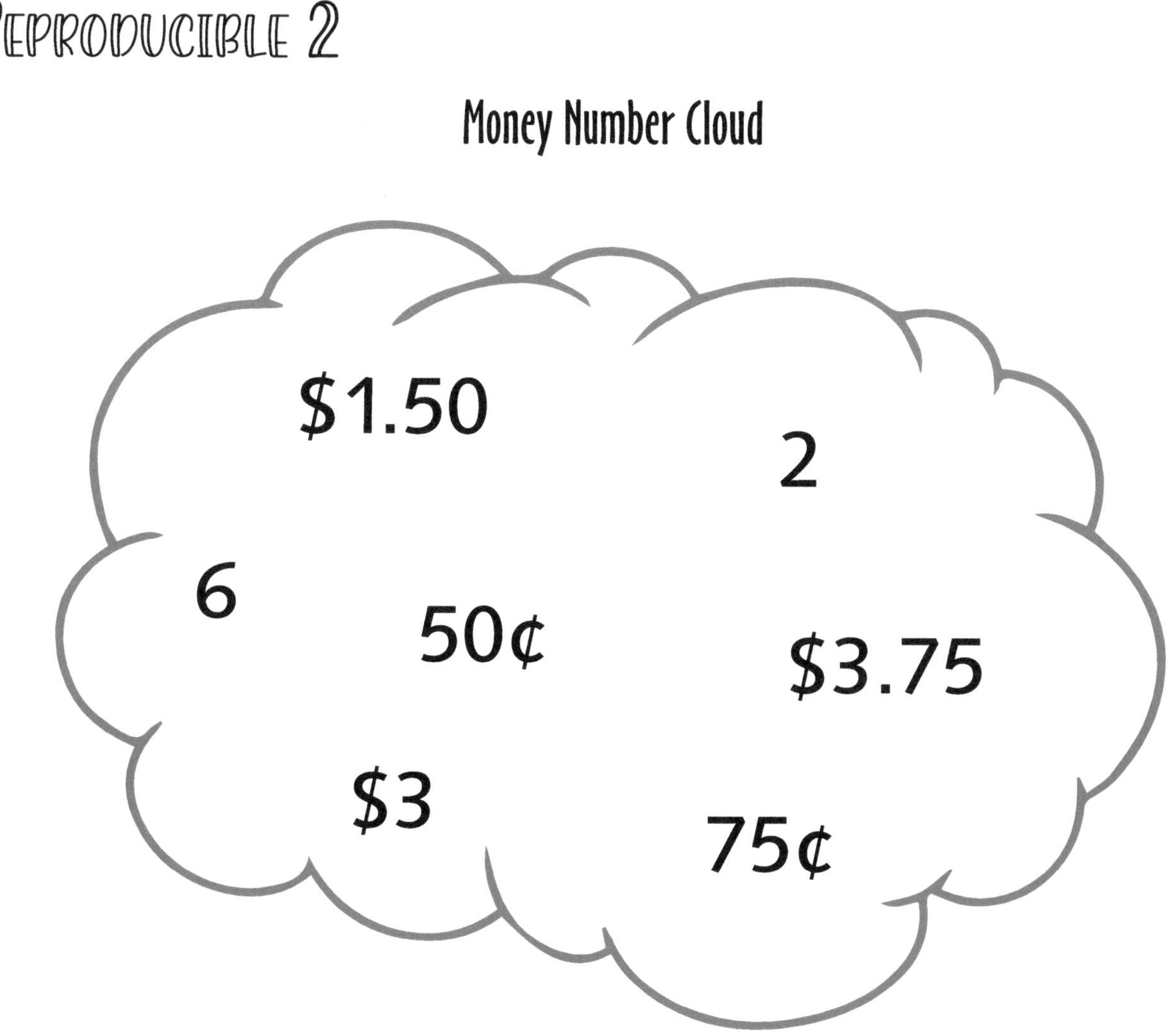

For use with Chapter 4, Question 3, section Grades 3–5: Money.

Reproducible 3

Jigsaw Piece

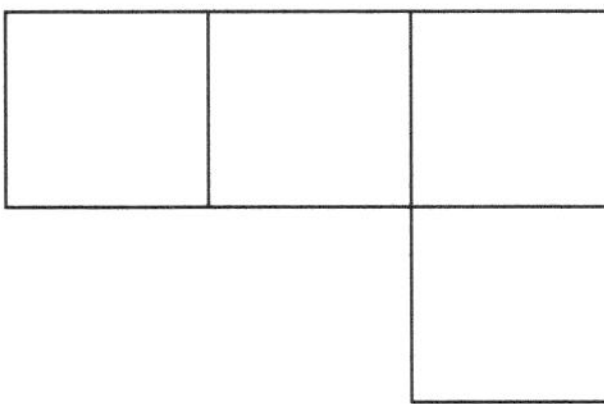

For use with Chapter 5, Question 3, section Grades K–2: Place Value.

Reproducible 4

Decimal Number Cloud

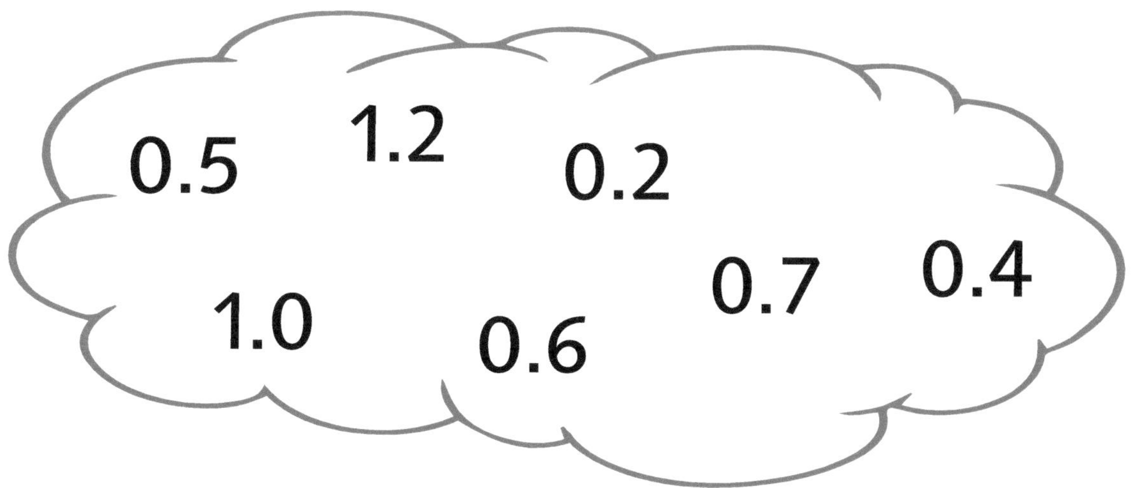

For use with Chapter 6, Question 27, section Grades 3–5: Decimal Concepts.

Reproducible 5

Triangles

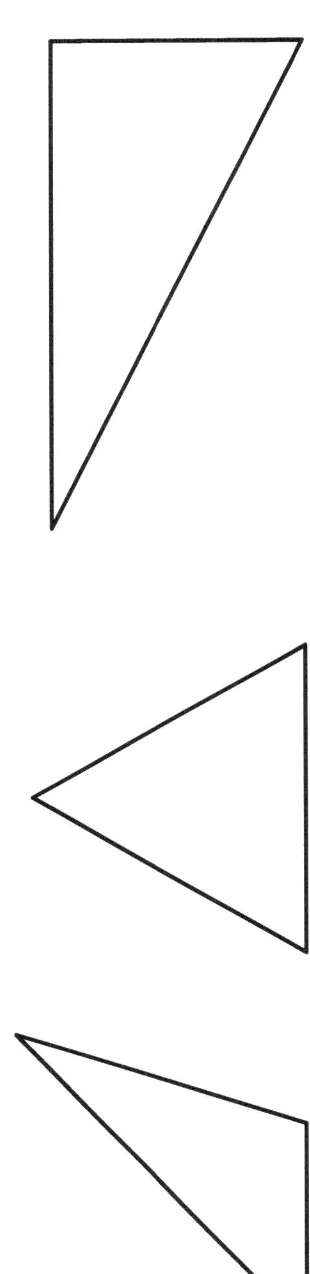

For use with Chapter 9, Question 2, section Grades K–2: Two-Dimensional Shapes.

Reproducible 6

Mixed Two-Dimensional Shapes

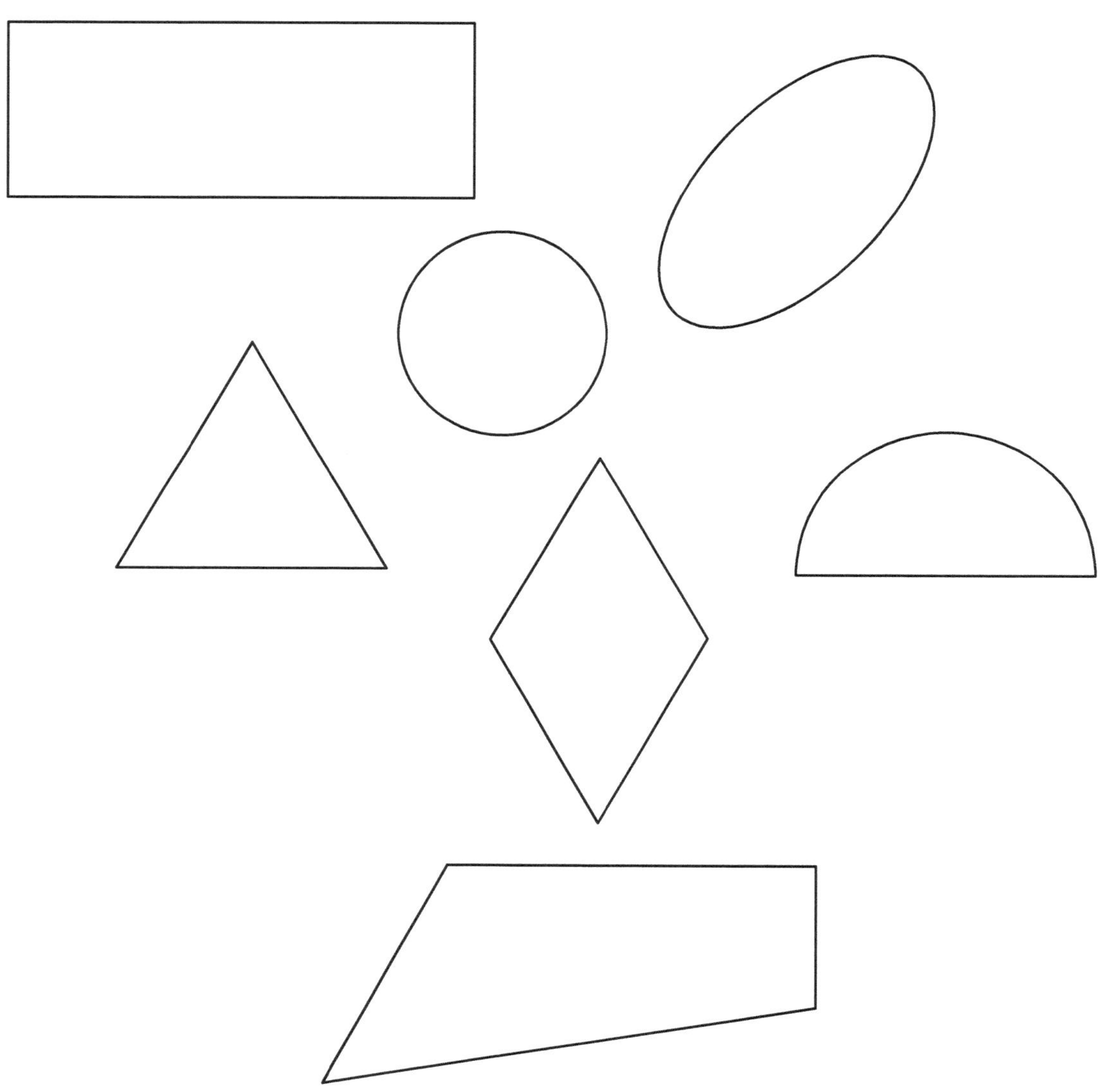

For use with Chapter 9, Question 5, section Grades K–2: Two-Dimensional Shapes.

Reproducible 7

Mixed Three-Dimensional Shapes

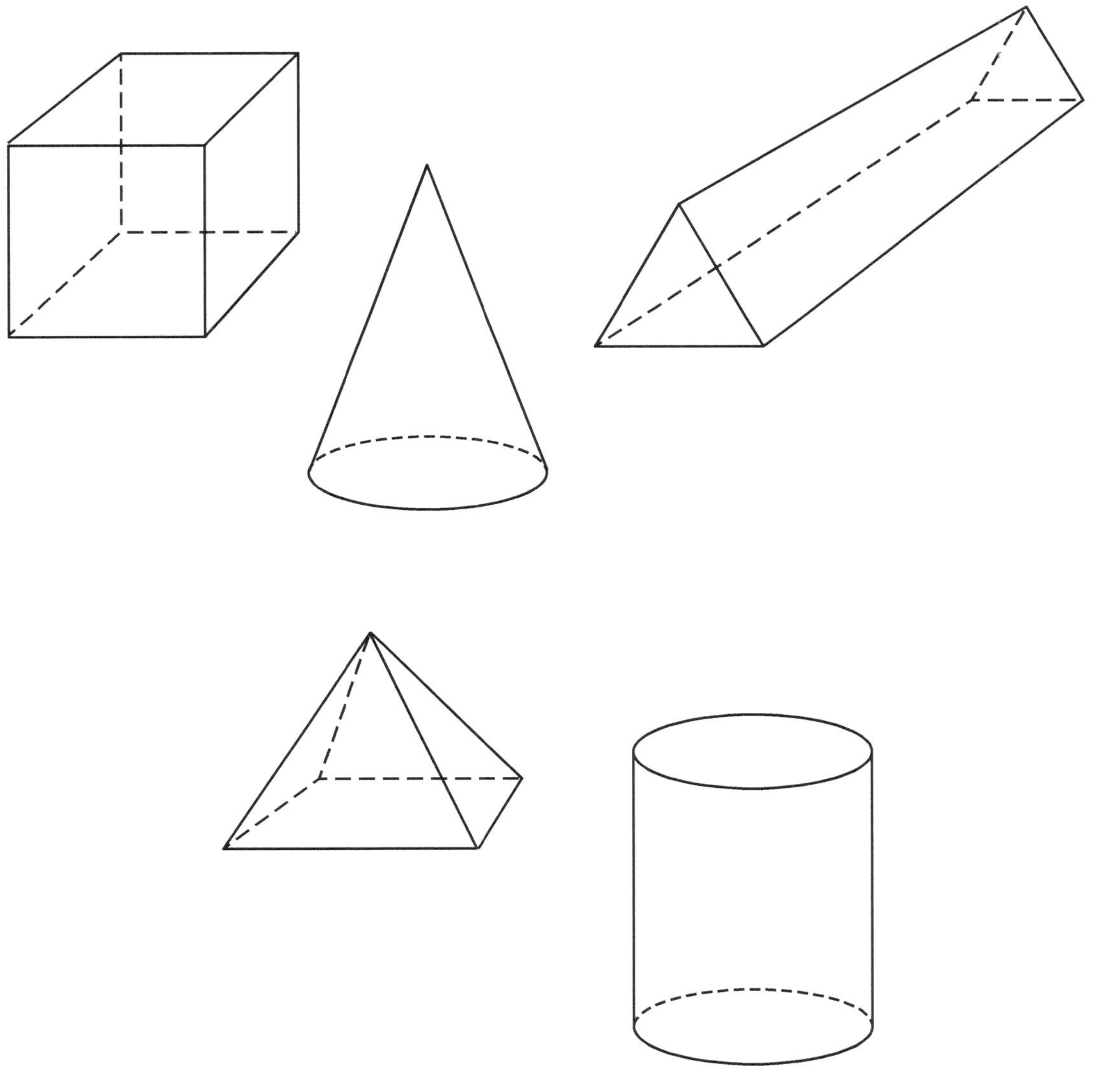

For use with Chapter 9, Question 12, section Grades K–2: Three-Dimensional Shapes.

Reproducible 8

Walking Path Shape

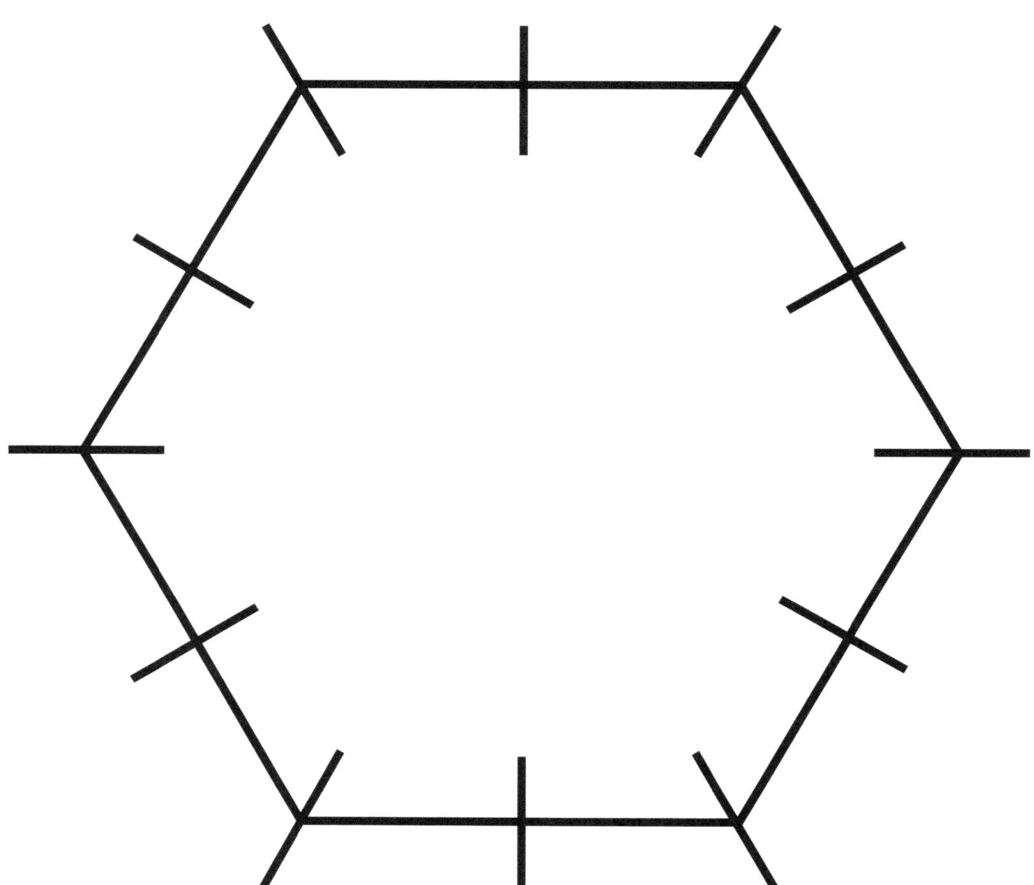

For use with Chapter 9, Question 7, section Grades 3–5: Two-Dimensional Shapes.

Reproducible 9

Block Diagram

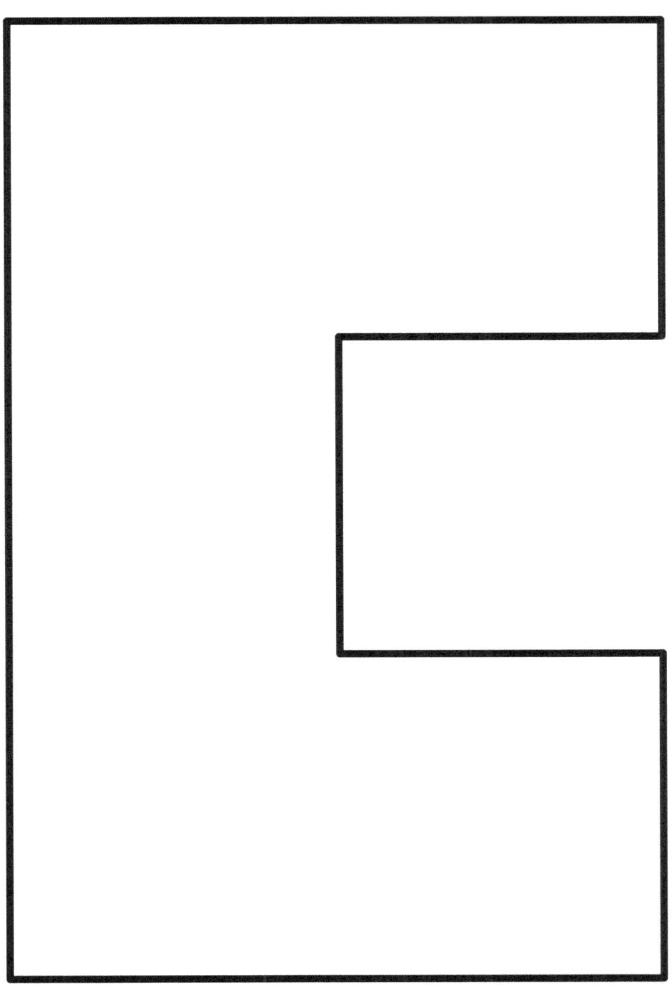

For use with Chapter 9, Question 2, section Grades 3–5: Three-Dimensional Shapes.

Reproducible 10

Mystery Bar Graph

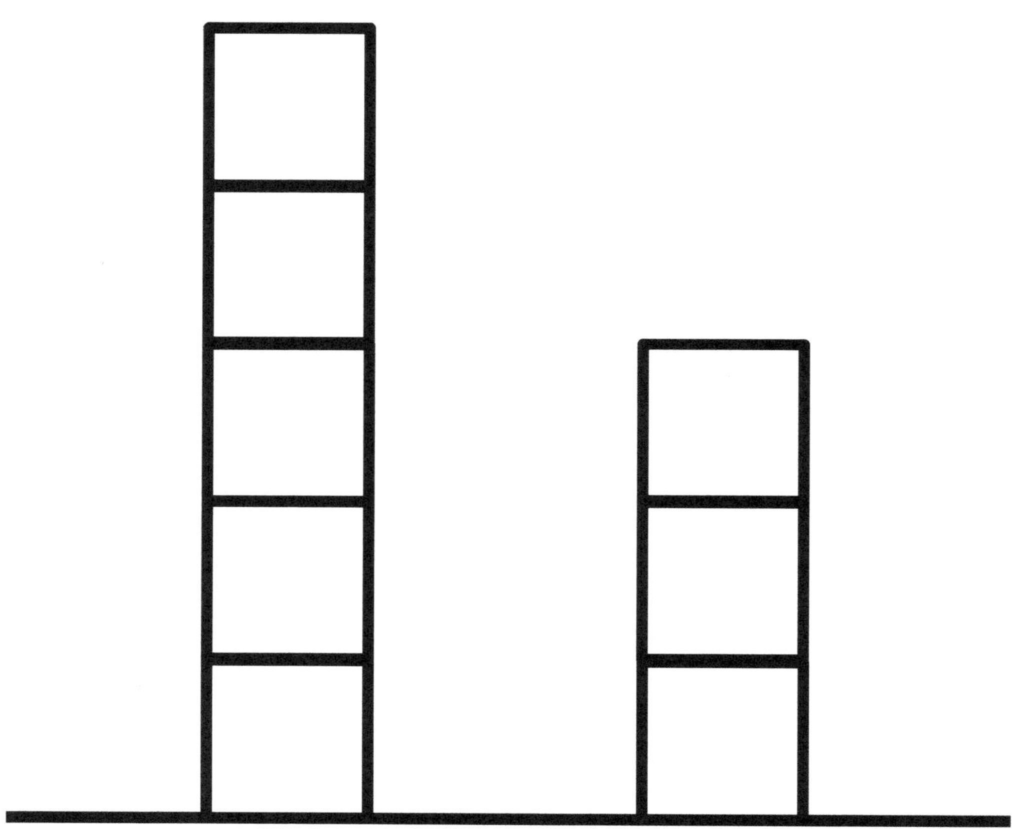

For use with Chapter 10, Question 1, section Grades K–2: Data.

Reproducible 11

Mystery Line Plot

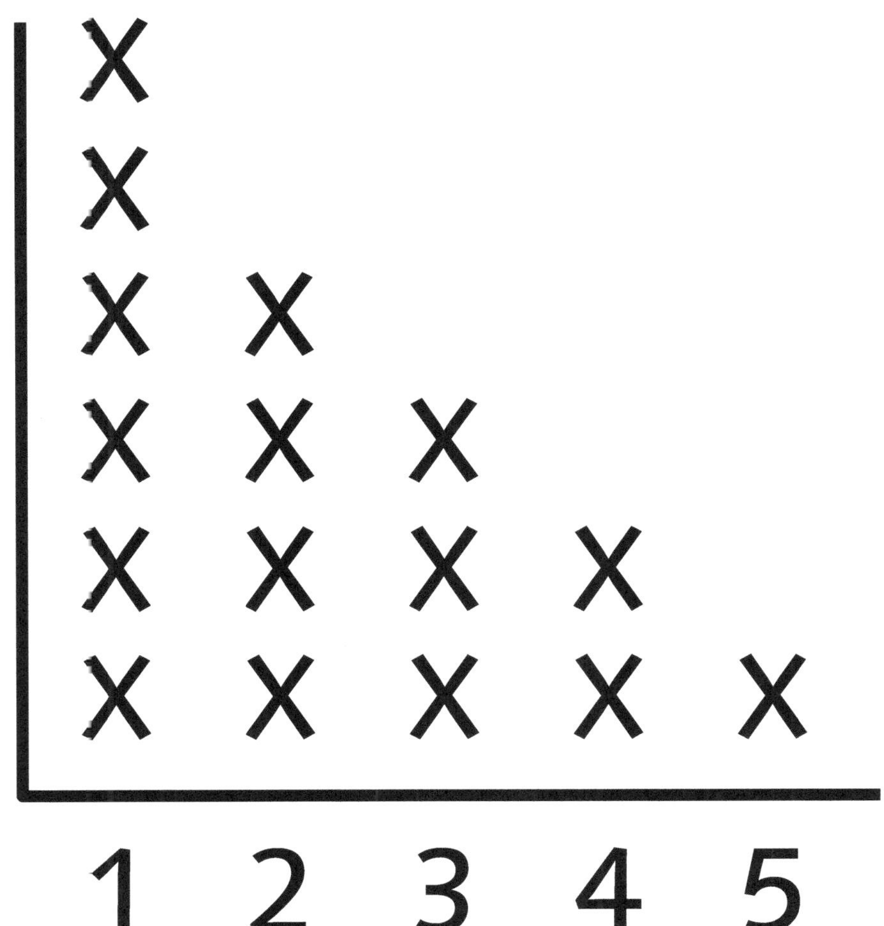

For use with Chapter 10, Question 5, section Grades K–2: Data.

Reproducible 12

Blank Spinner Faces

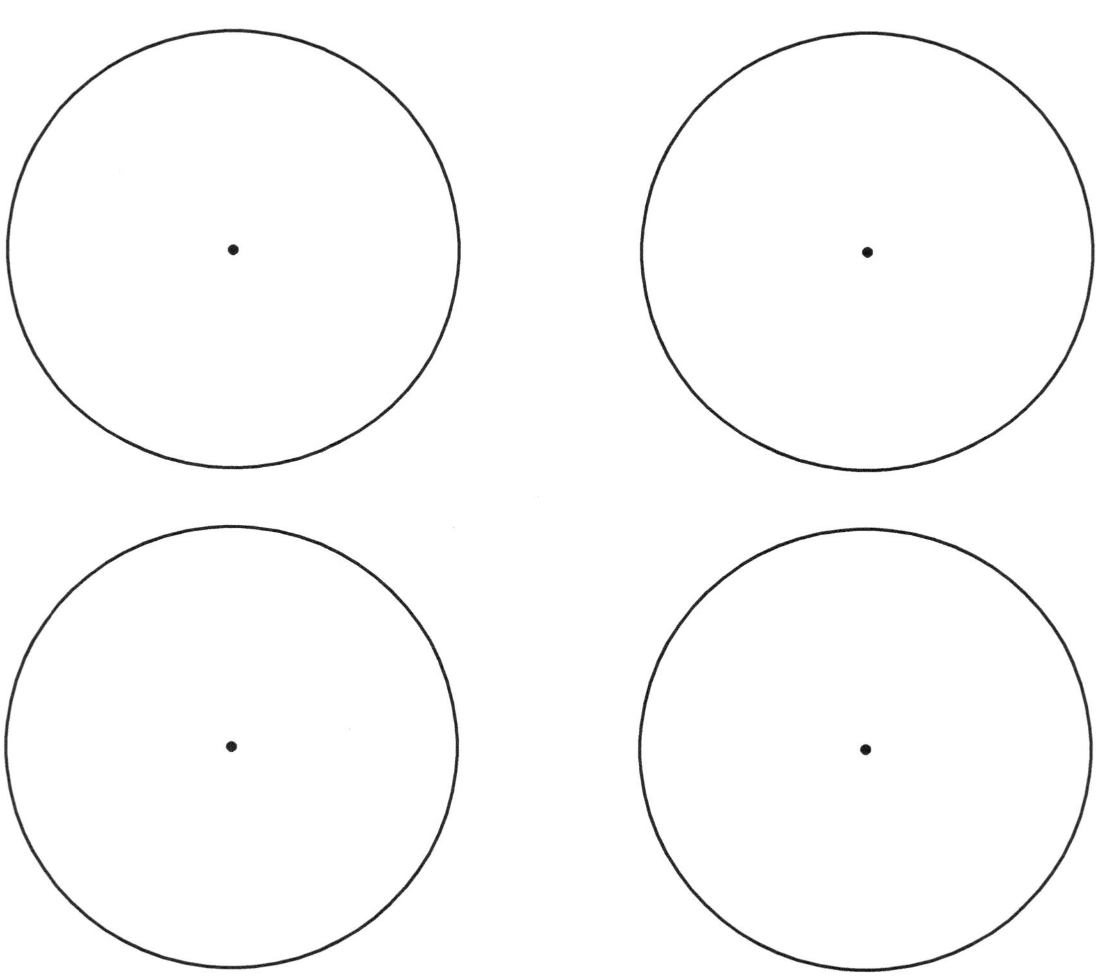

For use with Chapter 10, Question 7, section Grades 3–5: Chance.

Reproducible 13

Favorite Shows Pie Graph

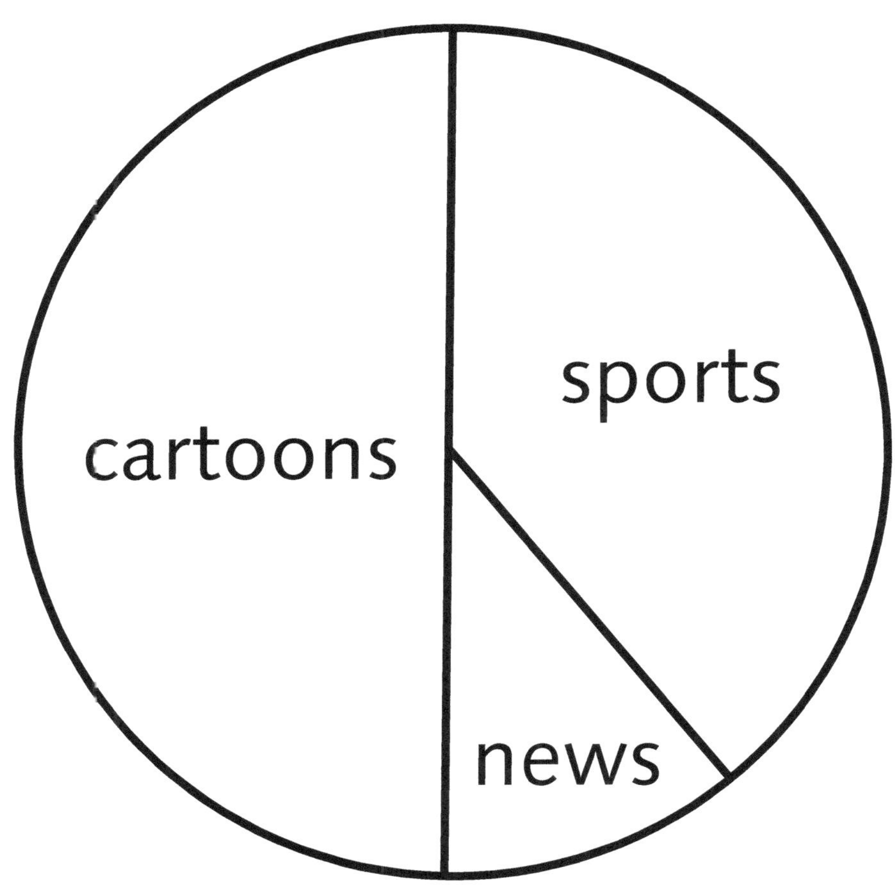

For use with Chapter 10, Question 1, section Grades 3–5: Data.

Reproducible 14

Mystery Pictograph

For use with Chapter 10, Question 2, section Grades 3–5: Data.

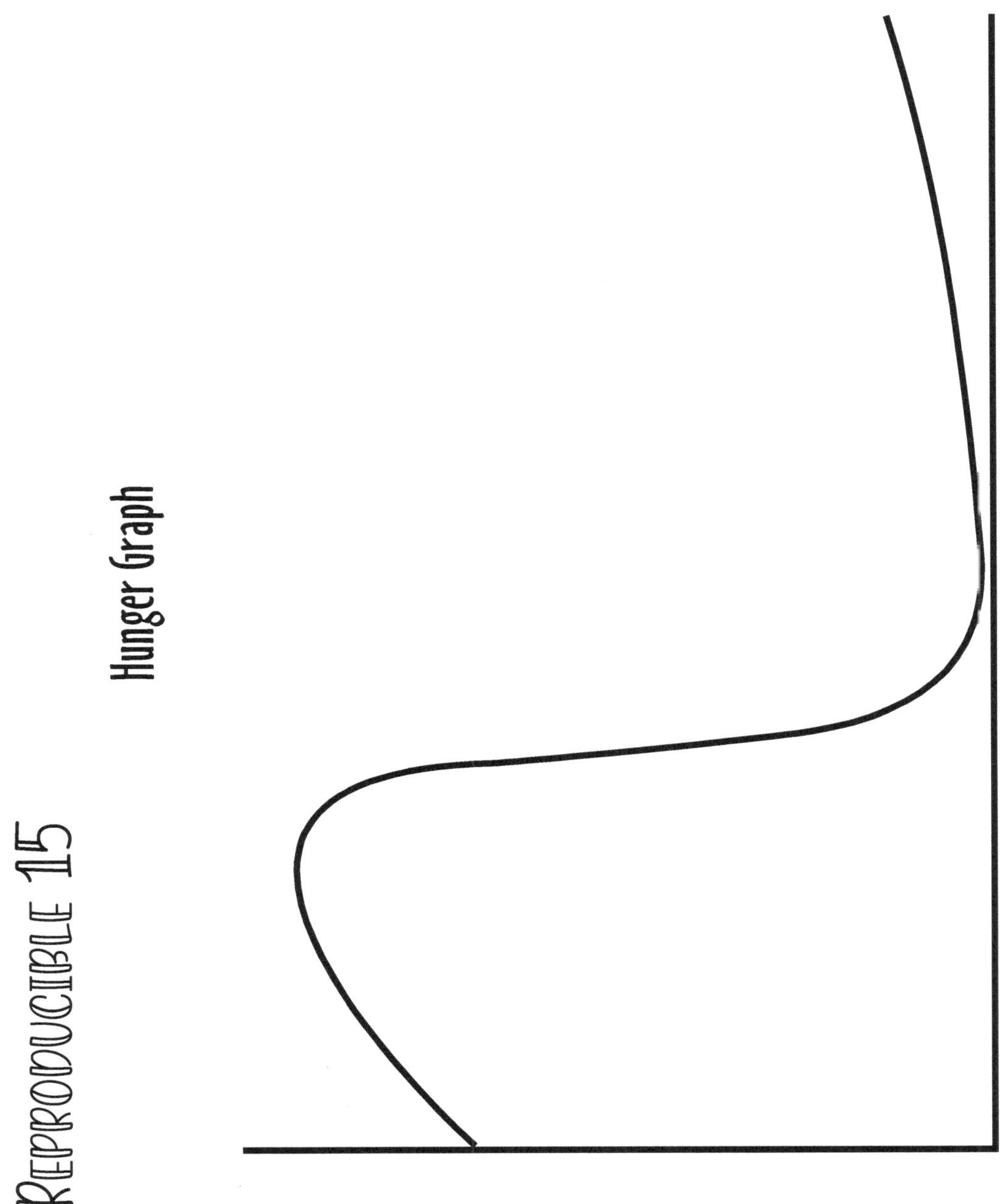

For use with Chapter 10, Question 3, section Grades 3–5: Data.

Reproducible 16

Mystery Survey Results

Place	Tally
library	///
playground	~~////~~ ~~////~~ //
under the trees	~~////~~ //
driveway	///

For use with Chapter 10, Question 4, section Grades 3–5: Data.

From *Good Questions for Math Teaching, Grades K–5, Second Edition,* by Peter Sullivan and Pat Lilburn. Portsmouth, NH: Heinemann.

Reproducible 17

Mystery Pie Chart

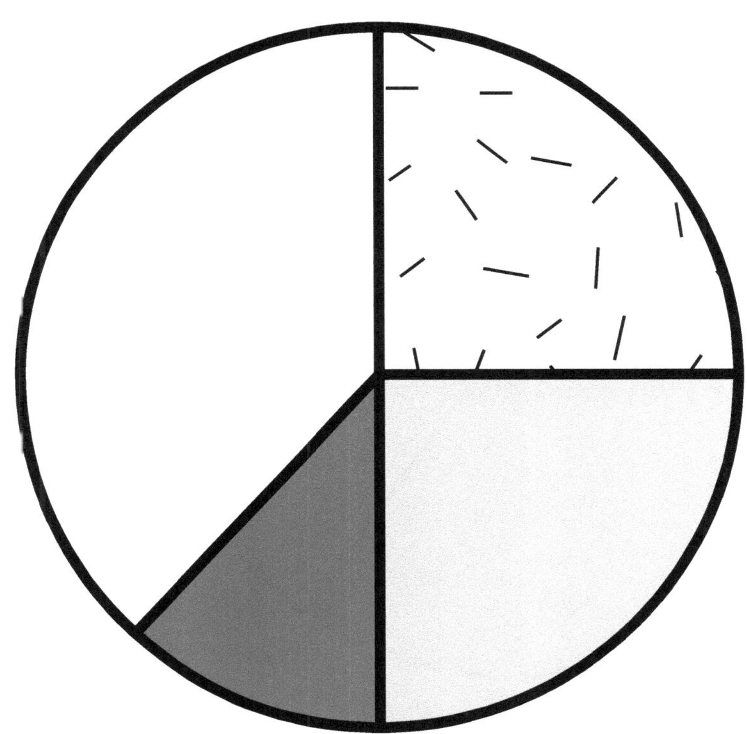

For use with Chapter 10, Question 7, section Grades 3–5: Data.

Reproducible 18

Sports Pie Chart

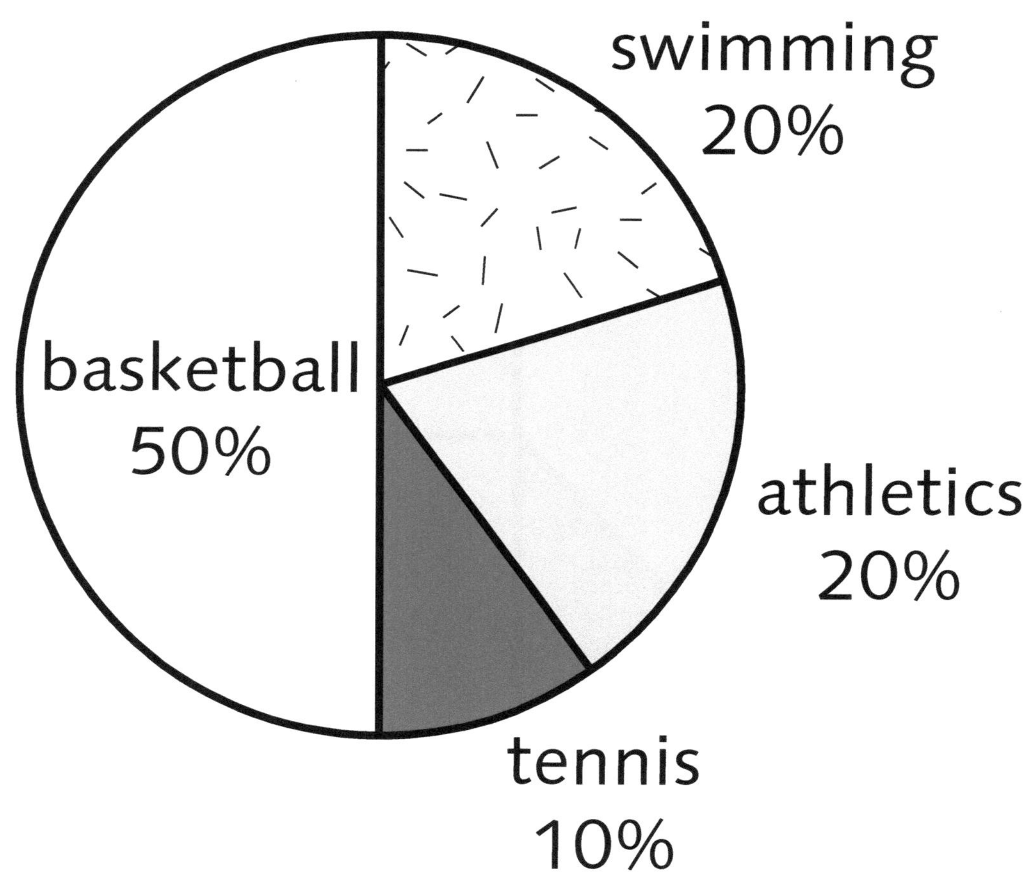

For use with Chapter 10, Question 8, section Grades 3–5: Data.

Reproducible 19

Animal Pictograph

For use with Chapter 10, Question 9, section Grades 3–5: Data.

Reproducible 20

Measurement Line Plot

							X				
					X		X				
				X	X		X				
				X	X		X				
	X	X		X	X	X	X				
	X	X		X	X	X	X			X	
8	$8\frac{1}{4}$	$8\frac{1}{2}$	$8\frac{3}{4}$	9	$9\frac{1}{4}$	$9\frac{1}{2}$	$9\frac{3}{4}$	10	$10\frac{1}{4}$	$10\frac{1}{2}$	$10\frac{3}{4}$

For use with Chapter 10, Question 10, section Grades 3–5: Data.

Reproducible 21

Graph of Children Talking

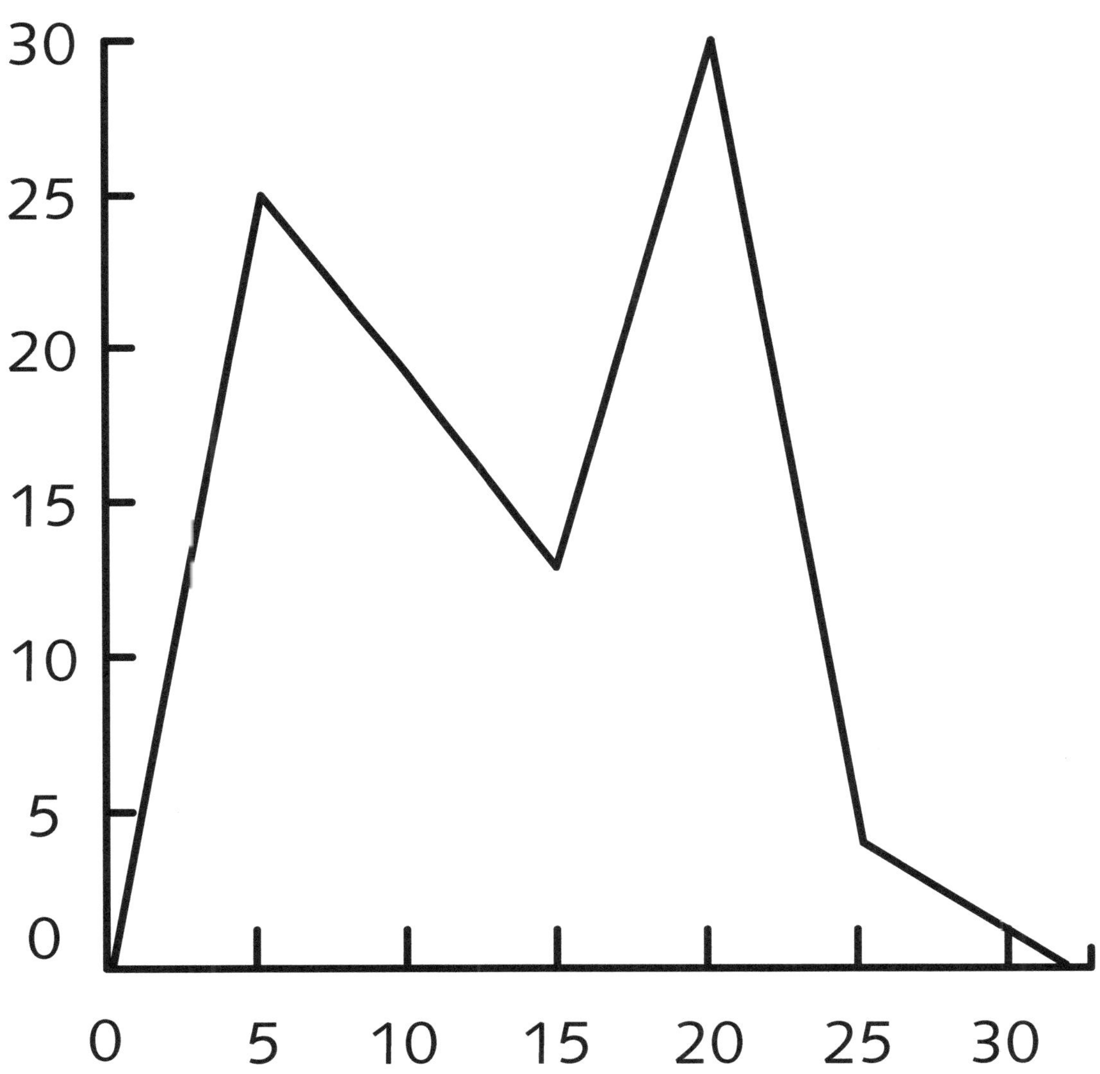

For use with Chapter 10, Question 11, section Grades 3–5: Data.

Reproducible 22

Cubic Structure

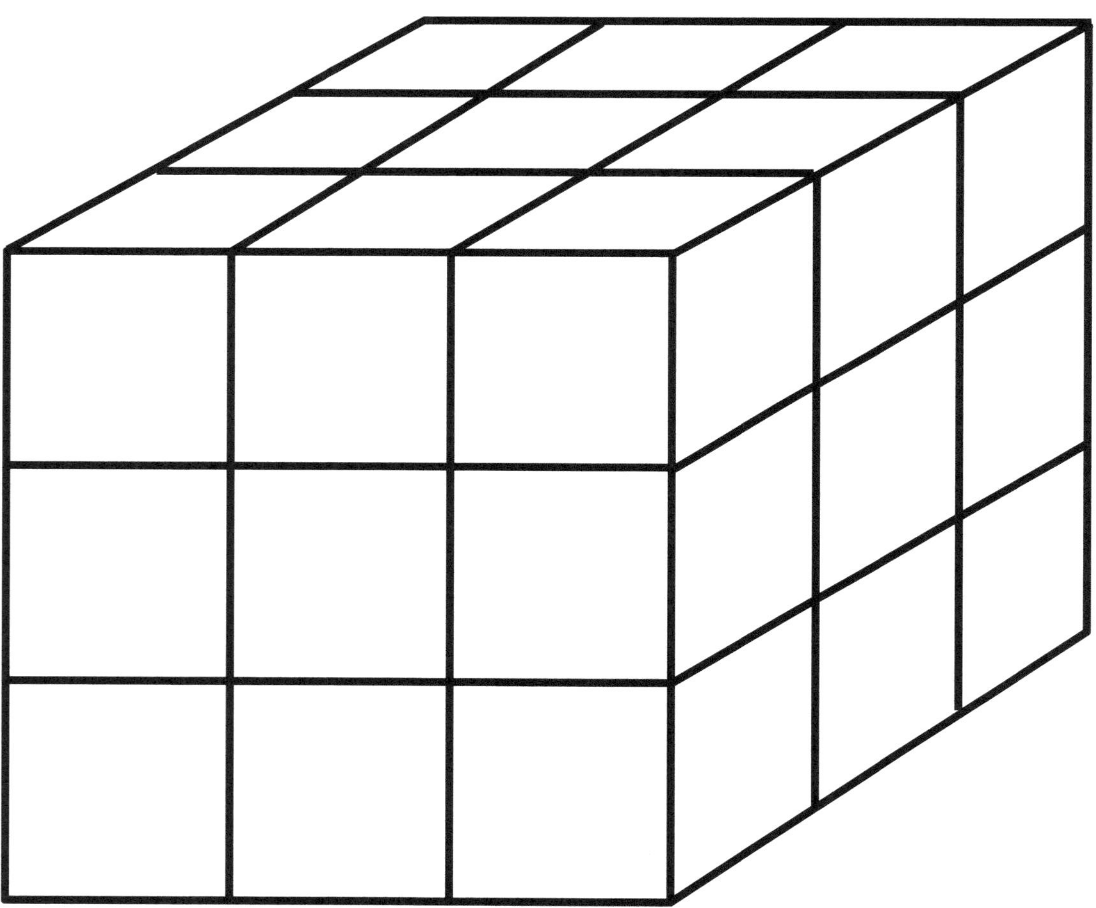

For use with Chapter 11, Question 5, section Grades 3–5: Volume.

Reproducible 23

Rectangular Box

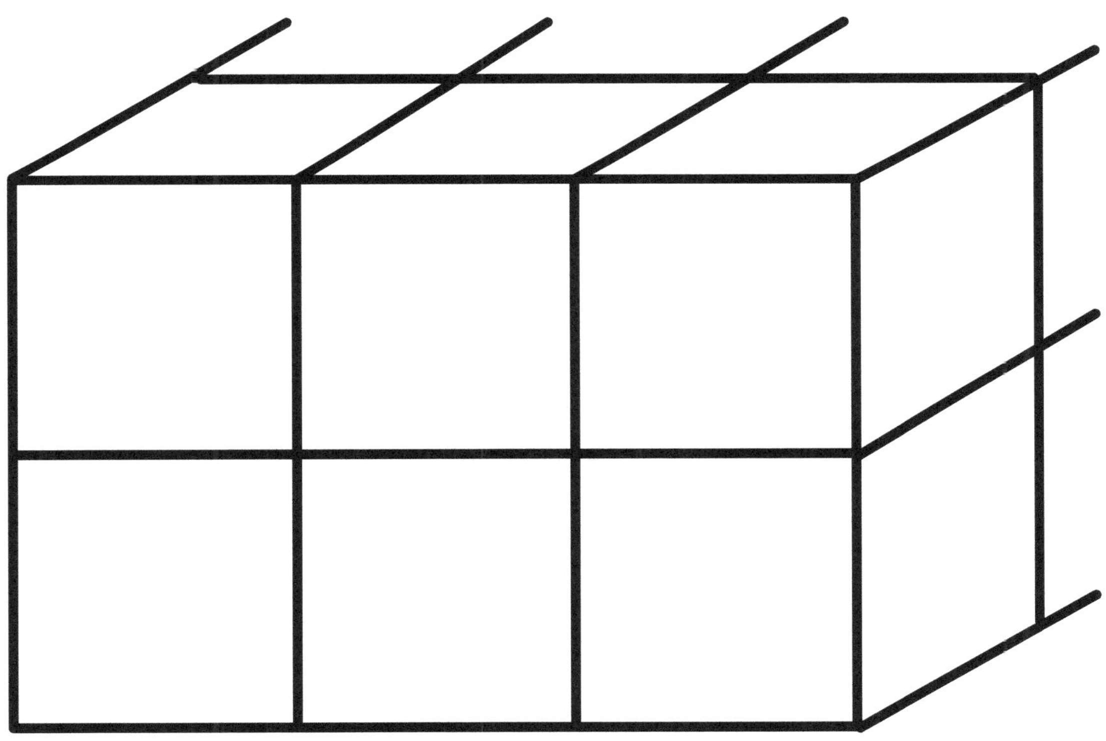

For use with Chapter 11, Question 6, section Grades 3–5: Volume.

Reproducible 24

Cube Diagram

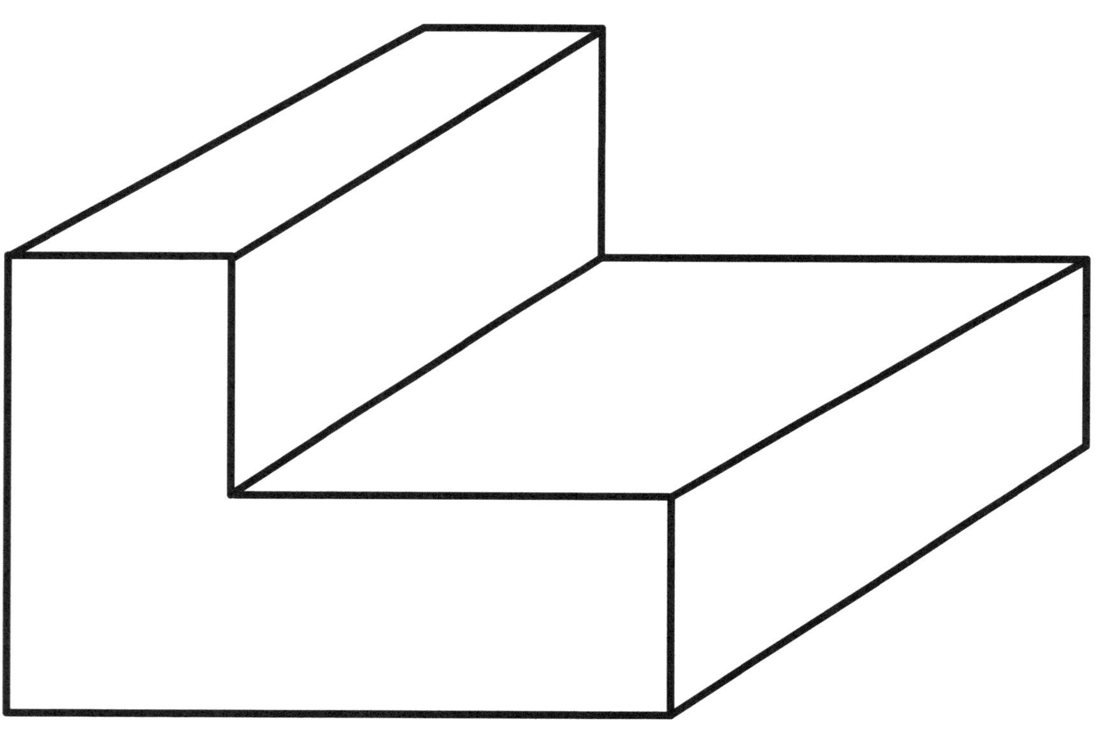

For use with Chapter 11, Question 10, section Grades 3–5: Volume.

Reproducible 25

Letter O

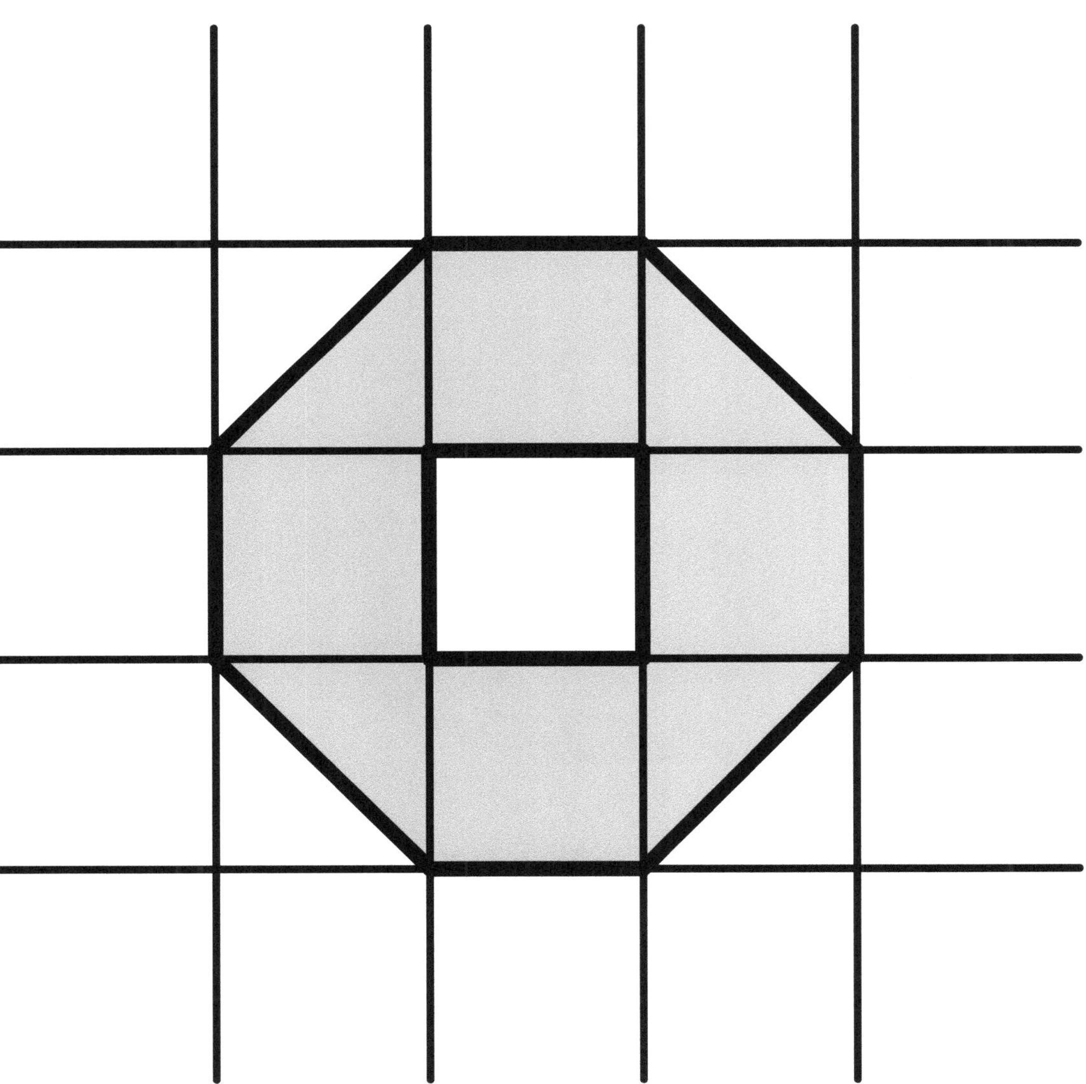

For use with Chapter 13, Question 13, section Grades 3–5: Area.

Reproducible 26

Package with Ribbon

For use with Chapter 11, Question 7, section Grades 3–5: Length and Perimeter.

Reproducible A

Hundreds Chart

1	2	3	4	5	6	7	8	9	10
11	12	13	14	15	16	17	18	19	20
21	22	23	24	25	26	27	28	29	30
31	32	33	34	35	36	37	38	39	40
41	42	43	44	45	46	47	48	49	50
51	52	53	54	55	56	57	58	59	60
61	62	63	64	65	66	67	68	69	70
71	72	73	74	75	76	77	78	79	80
81	82	83	84	85	86	87	88	89	90
91	92	93	94	95	96	97	98	99	100

For use with multiple questions as you see fit.

References

Chapin, Suzanne H., Catherine O'Connor, and Nancy Canavan Anderson. 2013. *Talk Moves: A Teacher's Guide for Using Classroom Discussions in Math, Grades K–6*. 3d ed. Sausalito, CA: Math Solutions.

Costa, Arthur L., and Bena Kallick, eds. 2000. *Activating and Engaging Habits of Mind*. Alexandria, VA: Association for Supervision and Curriculum Development.

Dantonio, Marylou, and Paul C. Beisenherz. 2001. *Learning to Question, Questioning to Learn: Developing Effective Teacher Questioning Practices*. Needham Heights, MA: Allyn and Bacon.